T0324489

A Quick Introduction to
Complex Analysis

A Quick Introduction to Complex Analysis

Kalyan Chakraborty
Harish-Chandra Research Institute, India

Shigeru Kanemitsu
Kindai University, Japan

Takako Kuzumaki
Gifu University, Japan

NEW JERSEY • LONDON • SINGAPORE • BEIJING • SHANGHAI • HONG KONG • TAIPEI • CHENNAI • TOKYO

Published by

World Scientific Publishing Co. Pte. Ltd.
5 Toh Tuck Link, Singapore 596224
USA office: 27 Warren Street, Suite 401-402, Hackensack, NJ 07601
UK office: 57 Shelton Street, Covent Garden, London WC2H 9HE

Library of Congress Cataloging-in-Publication Data
Names: Chakraborty, Kalyan. | Kanemitsu, Shigeru. | Kuzumaki, Takako, 1960–
Title: A quick introduction to complex analysis / by Kalyan Chakraborty
(Harish-Chandra Research Institute, India), Shigeru Kanemitsu (Kinki University, Japan),
Takako Kuzumaki (Gifu University, Japan).
Description: New Jersey : World Scientific, 2016. | Includes bibliographical references and index.
Identifiers: LCCN 2016023261| ISBN 9789813108509 (hardcover : alk. paper) |
ISBN 9789813108516 (pbk. : alk. paper)
Subjects: LCSH: Mathematical analysis--Textbooks.
Classification: LCC QA300 .C43 2016 | DDC 515/.9--dc23
LC record available at https://lccn.loc.gov/2016023261

British Library Cataloguing-in-Publication Data
A catalogue record for this book is available from the British Library.

Printed in Singapore

To the memory of Professor Dr. Katsumi Shiratani
April 7, 1932-December 28, 2004.
Supervisor of the 2nd and 3rd authors, an affectionate, sometimes stubborn, father.

Preface

This is a book on complex analysis written in a very reader-friendly way. It consists of two parts: Chapter 1—a quick introduction to complex analysis with applications and Chapter 2—applicable real and complex functions. In the applied sections of Chapter 1, the reader can gather some basic knowledge on fluid dynamics, thermodynamics, electrical circuits, (robust) control theory, and signal transmission. Also, in Chapter 2, it gives a smooth bridging from real analysis (those sections marked with *) to complex analysis, akin to **a bridge over troubled water**. The real analysis part can be skipped if the reader is already familiar with the contents. The book is meant for multiple purposes. It can be used for self-study, or as a text for courses of various length. For a short course, a quick introduction may be taught in a week or so. Skipping the real analysis part or application part, the main body may be taught in one semester. The book's reader-friendly style is manifested by plenty of worked-out examples and many exercises which can be used as problems in term-exams. It gives a chance for the reader to "read between lines", which is not common in mathematics books. The book therefore may be used by students in engineering disciplines, too. It has many unique features, like analytic continuation from real to complex functions, symbolically $x \to z$, in the sense that from formulas in real analysis one can warp immediately to those in complex analysis. One will master the use of partial fraction expansions to solve the DE by the method of Laplace transforms. If one follows some of the examples, then one can with ease master the method. Even a freshman can follow easily since we have included the calculus version. Even if one just goes through Chapter 1, one will have a feel of the theory. The important thing at the beginner's level is to get some feeling—meta knowledge—of what the theory is about.

Like field theory is for decomposing polynomials, function theory is for doing analysis on the complex plane, with more ease than in real calculus. Differentiation bears a strong resemblance to real analysis and only integration is cumbersome. The reader can learn nice techniques of residue calculus quite easily and get a sound basis of complex functions, which is often insufficient in engineering disciplines. Especially, the meaning is made clear for Fourier transforms via residue calculus for signal transmission and control theory.

An interested reader may go on further to reading more advanced books on complex functions. There are plenty of them. For example, [Ahlfors (1979)], [Apostol (1957)], [Dienes (1931)], [Gel'fond (1971)], [Lindelöf (1905)], [Rudin (1986)], [Titchmarsh (1939)], where each of them is quite readable and beneficial. Save for Gel'fond and Lindelöf which are devoted to residue calculus, others are authoritative accounts of complex function theory. Apostol, Rudin and Titchmarsh contain other material on real analysis, too. Gel'fond and Lindelöf can be read with [Mitrinović and Kečkić (1984)]. The books [Karatsuba (1995)] and [Tatuzawa (1980)] are rather focused on number-theoretic applications (cf. also [Segal (1981)]). For special functions, [Whittaker and Watson (1927)] is authoritative although there are many new modern books including [Kanemitsu and Tsukada (2007)] which is centered around the gamma and zeta-functions.

The book's purpose is to provide you with meta knowledge and get you familiarized with complex analysis more smoothly and light-heartedly than other books. If you are in a real hurry, just go through §1.1 and you will be ready in three days.

We would like to express our hearty thanks to Dr. Jay Mehta for his devoted help toward the completion of the book. Thanks are due to Ms. Kwong Lai Fun for her patience and encouragement.

Cover picture: An image of Hiroshige on the country of Echigo

Kalyan Chakraborty, Shigeru Kanemitsu and Takako Kuzumaki

Contents

Chapter 1

A Quick Introduction to Complex Analysis with Applications

1.1 The quickest introduction to complex analysis

This section serves as a review or survey material for those who are in a hurry to recall what the main ingredients in complex analysis are **in three days**. It presupposes some basic knowledge of complex numbers and functions etc. If you encounter something whose meaning is not clear to you, you must make a recourse to a more standard presentation of the theory in the subsequent sections.

Since complex function theory is in a sense infinitesimal calculus of complex-valued functions in the complex variable, it follows that the most fundamental ingredients in the theory are **differentiation** and **integration**; the former of which we define in the same way as with real functions, i.e. the function $f(z)$ is **differentiable** at $z = z_0$ in its domain of definition D if

$$\lim_{z \to z_0} \frac{f(z) - f(z_0)}{z - z_0} \tag{1.1}$$

exists, in which case, the limit value is denoted by $f'(z_0)$ called the **derivative** of f at z_0. Formally this is the same as the derivative of a real function, the only difference being that the limit is taken in a planar domain $D \subset \mathbb{C}$.

On the other hand, the latter is more complicated since integration means a contour integral. However, the below-mentioned "performing the change of variable principle" makes it quite easily accessible.

Definition 1.1. We say that a complex function is **analytic** (or sometimes **holomorphic** or **regular**) in a **domain** if it is **differentiable** at each point of the domain, where differentiability means the existence of the derivative, and a domain (sometimes referred to as a **region**) mathematically means a connected set; we simply understand a domain to be a certain plane figure ($\subset \mathbb{C}$) with interior and with the boundary curve. We usually assume that domains are arc-wise connected. Typical domains are the rectangles (parallelopipeds) and circles and there is no need to worry about what domains are.

We assume throughout that a curve is a **piecewise smooth (Jordan) curve** described by the parametric expression

$$z = z(t) = x(t) + iy(t), \quad t \in [a, b]. \tag{1.2}$$

E.g., the unit circle $C : |z| = 1$ is given by

$$z = z(t) = e^{2\pi it}, \quad t \in [0, 1] \tag{1.3}$$

or by

$$C : z = z(t) = e^{it}, \ t \in [0, 2\pi],$$

where the complex exponential function e^{it}, cf. §1.3.2.

By the Jordan curve theorem, such a curve encircles a domain D. In this context, we denote the curve by ∂D and refer to it as its boundary.

Note that this is a **positively-oriented curve** to the effect that if you traverse the curve, you'll see the inside on your left. We assume all the curves are positively oriented. The expression (1.2) for the unit circle is in counter-clockwise direction. The same curve oriented in the opposite direction, denoted $-C$, is given by

$$-C : z = z(t) = e^{2\pi i(1-t)}, \ t \in [0, 1].$$

Remark 1.1. The notion of analytic functions is inseparably connected with their domains (which are connected). Indeed, an analytic function $f(z)$ and its domain D of analyticity is to be considered as a pair (f, D). Unlike calculus, we do not consider the case of differentiability at one point but on a domain and we say that a function is **analytic at a point** or **on a curve** if it is analytic in a domain containing the point or the curve.

There are two main streams of ideas in complex analysis. One is due to Weierstrass and appeals to the power series expansion. The other is due to Cauchy and describes the analyticity in terms of contour integrals. Complex integrals are all line (or contour) integrals and are slightly different from the ordinary Riemann integrals $\int_a^b f(x)\,dx$. However we may think of it as **performing the change of variable** $dz = z'(t)\,dt$:

$$\int_C f(z)\,dz = \int_a^b f(z(t))z'(t)\,dt \tag{1.4}$$

if the curve C is given by (1.2). We note that

$$\int_{-C} f(z)\,\mathrm{d}z = -\int_C f(z)\,\mathrm{d}z.$$

The **Weierstrass main theorem** which is equivalent to **Weierstrass double series theorem**, Corollary 2.2 reads

Theorem 1.1. (Weierstrass's main theorem) *A function $f(z)$ is analytic in a domain D if and only if it is expanded into a* **power series** *(Taylor series) at each point z_0 of D:*

$$f(z) = \sum_{n=0}^{\infty} a_n (z - z_0)^n. \tag{1.5}$$

If f is analytic, then it has derivatives of all orders and (1.5) *holds.*

The second part of the theorem, which is a feature very distinct from real analytic functions, is often referred to as **Goursat's theorem.**

This implies the **consistency** (or **unicity**) **theorem**

Theorem 1.2. (Consistency theorem) *If two functions are analytic in a domain D and they coincide on a subset S of D having an accumulation point in D, then they must coincide on the whole of D.*

In particular, if the domain D contains a segment of the real axis and the functions coincide on it, then they coincide on the whole of D.

The consistency theorem is a basis for the **principle of analytic continuation** to the effect that

Theorem 1.3. (Analytic continuation) *If two analytic functions coincide e.g. on a segment of the real axis, then one is an analytic continuation of the other.*

Cf. the passages after Remark 1.2.

The following **Cauchy integral theorem** is the most fundamental in complex analysis and one of the most far-reaching theorem in whole mathematics.

Theorem 1.4. (Cauchy integral theorem) *A function $f(z)$ is analytic in a domain D if and only if for any closed curve C within D, we have*

$$\int_C f(z)\,\mathrm{d}z = 0. \tag{1.6}$$

Fig. 1.1 August Louis Cauchy (1789-1857)

Example 1.1. The circle with center at α and radius r, $|z-\alpha| = r$, can be expressed as

$$C: \ z = \alpha + re^{2\pi i t}, \quad t \in [0,1]. \tag{1.7}$$

We evaluate the integral

$$I = \int_C \frac{1}{z-\alpha} \, \mathrm{d}z.$$

Substituting (1.7), we get

$$I = \int_0^1 \frac{1}{re^{2\pi i t}} re^{2\pi i t} (2\pi i) \, \mathrm{d}t = 2\pi i.$$

How about $I_n = \int_C \frac{1}{(z-\alpha)^n} \, \mathrm{d}z, \ \ n \geq 2$? Similarly, we have

$$I_n = \int_0^1 \frac{re^{2\pi i t}}{(re^{2\pi i t})^n} 2\pi i \, \mathrm{d}t = \frac{2\pi i}{r^{n-1}} \int_0^1 e^{2\pi i (n-1)t} \, \mathrm{d}t = 0.$$

By the Cauchy integral theorem,

$$\int_C (z-\alpha)^n \, \mathrm{d}z = 0, \quad n \geq 0.$$

Hence we have the evaluation for any integer $n \in \mathbb{Z}$

$$\int_C (z-\alpha)^n \, \mathrm{d}z = \begin{cases} 0, & n \neq -1 \\ 2\pi i, & n = -1, \end{cases} \tag{1.8}$$

where C is a circle enclosing the point $z = \alpha$.

Contraction principle

If $f(z)$ is analytic in a domain D except for a point $z = \alpha \in D$, then we may reduce the integral along any curve $C \subset D$ containing α to a circle c containing α, i.e. **we may contract a curve to a circle and apply** (1.8) under the assumption that interchange of the limits is legitimate.

For we connect c and C by two lines L_1 and L_2 oriented in opposite way. The curve $C_1 = -c \cup L_1 \cup L_2 \cup C$ encircles a domain, where $f(z)$ is analytic, so that

$$0 = \int_{C_1} f(z)\,\mathrm{d}z = \int_{-c} + \int_{L_1} + \int_{L_2} + \int_C f(z)\,\mathrm{d}z,$$

which gives

$$\int_C f(z)\,\mathrm{d}z = \int_c f(z)\,\mathrm{d}z.$$

Thus we find that complex analysis is very geometric (topological).

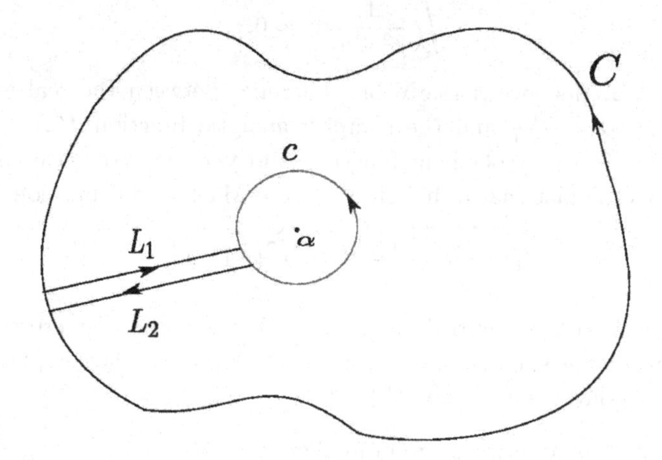

Fig. 1.2 Contraction to a circle

Example 1.2. Consider the function $f(z) = \frac{1}{z^2+1}$ which has simple poles at $z = \pm i$. A simple pole means that we have a denominator $z - i$ and $z + i$, (for more details, see below). Let

$$C_i : z - i = \frac{1}{2}e^{2\pi i t}, \quad t \in [0,1].$$

Then

$$\int_{C_i} f(z)\,\mathrm{d}z = \frac{1}{2i}\int_{C_i}\left(\frac{1}{z-i} - \frac{1}{z+i}\right)\mathrm{d}z,$$

where we applied the partial fraction expansion in §1.10.1. The first integral is already in (1.8) and is π, while for the second, if we substitute the parametric expression, then we are to face the integral

$$2\pi i \int_0^1 \frac{e^{2\pi it}}{e^{2\pi it}+4i}\,dt,$$

which is $2\pi i + 8\pi^2 \int_0^1 \frac{\cos 2\pi t - (\sin 2\pi t + 4)i}{8\sin 2\pi t + 17}\,dt$, and we don't want to go on.

We should apply (1.6) to conclude that it is 0. Hence

$$\int_{C_i} \frac{1}{z^2+1}\,\mathrm{d}z = \pi.$$

Similarly, if C_{-i} is the circle with center at $z=-i$ and with radius $\frac{1}{2}$, then

$$\int_{C_{-i}} \frac{1}{z^2+1}\,\mathrm{d}z = -\pi,$$

while for any circle C with center at the origin and radius $0<r<1$,

$$\int_C \frac{1}{z^2+1}\,\mathrm{d}z = 0.$$

Remark 1.2. Here we notice a big difference between the real analytic function $f(x) = \frac{1}{x^2+1}$ and the complex analytic function $f(z) = \frac{1}{z^2+1}$. Indeed, $f(x)$ is a very obedient function, and you may not have paid any attention to the fact that although it has the Maclaurin expansion

$$f(x) = \frac{1}{1-(-x^2)} = \sum_{n=0}^{\infty}(-1)^n x^{2n},$$

it is absolutely convergent **only** in $|x|<1$. Why is there such a restriction? It is because of the singularities at $z=\pm i$ which prevent the complex power series being convergent outside of $|z|<1$.

Weierstrass' double series theorem implies

Theorem 1.5. *If a function is analytic in a domain D containing a point α, then it can be expanded into a Taylor series around α in the maximum disc that is contained in D, and moreover, the real analytic series for $f(x)$ can be uniquely continued analytically to $f(z)$.*

Therefore inside the domain of analyticity, we may simply write z for x and get an analytic function which involves the real analytic function as a special case.

$$f(x) = \frac{1}{x^2+1} \to f(z) = \frac{1}{z^2+1},\ |z|<1.$$

1.2 Complex number system

This section is quite elementary and an advanced reader can skip it and go on to the next sections. We begin with an introduction of complex numbers since there is still a confusion arising from the meaning of these imaginary numbers which sound non-existing. But they do exist as plane vectors as we shall see.

We know that 2-dimensional real vectors (plane vectors) $\boldsymbol{z} = \begin{pmatrix} x \\ y \end{pmatrix} \in \mathbb{R}^2$ form a **vector space** under the addition (translation)

$$\boldsymbol{z} + \boldsymbol{z}' = \begin{pmatrix} x + x' \\ y + y' \end{pmatrix}, \quad \left(\boldsymbol{z}' = \begin{pmatrix} x' \\ y' \end{pmatrix}\right)$$

and the scalar multiplication ($c \in \mathbb{R}$)

$$c\boldsymbol{z} = \begin{pmatrix} cx \\ cy \end{pmatrix}.$$

There are multiplications defined:
scalar product:

$$\boldsymbol{z} \cdot \boldsymbol{z}' = xx' + yy' \in \mathbb{R},$$

vector product:

$$\boldsymbol{z} \times \boldsymbol{z}' = \begin{vmatrix} x & x' \\ y & y' \end{vmatrix} = xy' - x'y \in \mathbb{R},$$

where the middle term indicates the determinant.

Example 1.3. (The first construction) This will be reviewed in Exercise 2.17. We introduce the vector $\bar{\boldsymbol{z}} = \begin{pmatrix} x \\ -y \end{pmatrix}$, which is a reflection of $\boldsymbol{z}$ with respect to the x-axis, and combine these two multiplications in the following $*$-operation due to Gauss:

$$\boldsymbol{z} * \boldsymbol{z}' = \begin{pmatrix} \bar{\boldsymbol{z}} \cdot \boldsymbol{z}' \\ \bar{\boldsymbol{z}} \times \boldsymbol{z}' \end{pmatrix} = \begin{pmatrix} xx' - yy' \\ xy' + x'y \end{pmatrix}.$$

E.g. if we label $\begin{pmatrix} 0 \\ 1 \end{pmatrix}$ by i, then we have

$$i^2 = i * i = \begin{pmatrix} 0 \\ 1 \end{pmatrix} * \begin{pmatrix} 0 \\ 1 \end{pmatrix} = \begin{pmatrix} -1 \\ 0 \end{pmatrix}, \quad i^2 = -1.$$

In view of

$$\boldsymbol{z} = \begin{pmatrix} x \\ y \end{pmatrix} = x \begin{pmatrix} 1 \\ 0 \end{pmatrix} + y \begin{pmatrix} 0 \\ 1 \end{pmatrix} = x\boldsymbol{e}_1 + yi,$$

where $\boldsymbol{e}_1 = \begin{pmatrix} 1 \\ 0 \end{pmatrix}$, we think of the vector $\boldsymbol{z}$ as a "number", a **complex number**, $z = x + iy$:

$$\mathbb{R}^2 \ni \boldsymbol{z} = \begin{pmatrix} x \\ y \end{pmatrix} \longleftrightarrow z = x + iy \in \mathbb{C}. \tag{1.9}$$

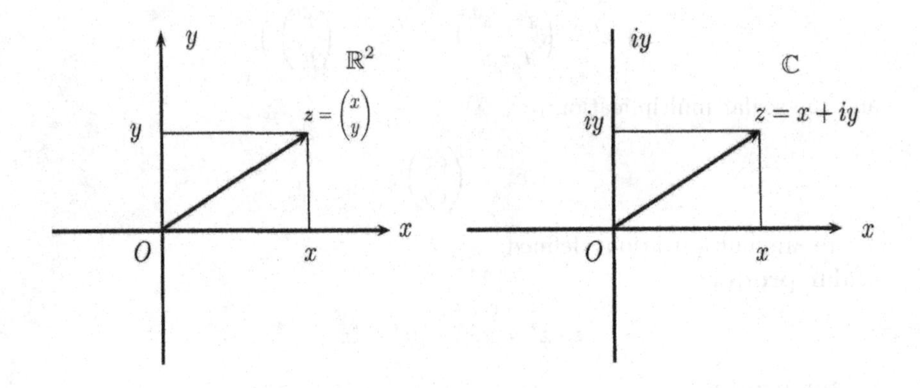

Fig. 1.3 Realization of $\mathbb{C}$ as $\mathbb{R}^2$

Since we just denote the basis vector $\boldsymbol{e}_2 = \begin{pmatrix} 0 \\ 1 \end{pmatrix}$ as i, the above correspondence is $1:1$ and moreover, it turns out that the star product is the same as ordinary multiplication of numbers with i replaced by -1:

$$zz' = z \cdot z' = (x + iy)(x' + iy') = xx' - yy' + i(xy' + x'y) \longleftrightarrow \boldsymbol{z} * \boldsymbol{z}'.$$

We can easily prove that the system $(\mathbb{R}^2, +, *)$ forms a field, which we denote by $\mathbb{C} = (\mathbb{C}, +, \cdot)$ and refer to it as the **field of complex numbers**. In electrical engineering, it is customary to denote i by j because i is kept for current (cf. §1.11.1).

Proof. The only point is the division and here is a proof. $\boldsymbol{z} = \begin{pmatrix} x \\ y \end{pmatrix} = \boldsymbol{o} = \begin{pmatrix} 0 \\ 0 \end{pmatrix}$ if and only if $\begin{pmatrix} x \\ y \end{pmatrix} = \begin{pmatrix} 0 \\ 0 \end{pmatrix}$, i.e. if and only if $\boldsymbol{z} = \boldsymbol{o}$, i.e. if and only if $|\boldsymbol{z}| = \sqrt{x^2+y^2} = 0$. Hence for each $\boldsymbol{z} \neq \boldsymbol{o}$, there exists the inverse element $\boldsymbol{z}^{-1} = \frac{1}{x^2+y^2}\begin{pmatrix} x \\ -y \end{pmatrix} = \frac{1}{|\boldsymbol{z}|^2}\bar{\boldsymbol{z}}$, where $\bar{\boldsymbol{z}} = \begin{pmatrix} x \\ -y \end{pmatrix}$, so that $\boldsymbol{z}^{-1} * \boldsymbol{z} = \frac{\boldsymbol{z}\bar{\boldsymbol{z}}}{|\boldsymbol{z}|^2} = 1$.

We recall the length (norm, absolute value) of a vector

$$|\boldsymbol{z}| = \sqrt{x^2+y^2}$$

(the Pythagoras theorem).

By (1.9) we introduce the **absolute value** of the complex number $z = x + iy$ by

$$|z| = |x+iy| = |\boldsymbol{z}| = \sqrt{x^2+y^2}.$$

Don't do this: $|1+3i| = \sqrt{1+(3i)^2} = \sqrt{-8}$.

We may of course define the **distance** between two numbers z, z' by

$$d(z,z') = |z-z'| = \sqrt{(x-x')^2+(y-y')^2}.$$

For this to be a distance function it must satisfy the triangular inequality, which will be shown in Exercise 2.20, (iii).

Then $(\mathbb{C}, d)$ is a metric space which is **complete** because $\mathbb{R}^2$ is so. These are the simplest examples of the **Hilbert spaces**. Hence we may develop an analysis on it, **complex analysis**.

Example 1.4. (The second construction) Consider the 2-dimensional subspace of the 4-dimensional real vector space

$$M = \{\boldsymbol{z} = (x, -y, y, x) | x, y \in \mathbb{R}\} \subset \mathbb{R}^4$$

and introduce the componentwise addition (translation) and the new multiplication $\times$ for $\boldsymbol{z}_j = (x_j, -y_j, y_j, x_j), j = 1, 2$

$$\boldsymbol{z}_1 \times \boldsymbol{z}_2 = (x_1x_2 - y_1y_2, -(x_1y_2 + x_2y_1), x_1y_2 + x_2y_1, x_1x_2 - y_1y_2). \quad (1.10)$$

Then $(M, +, \times)$ forms a field isomorphic to $\mathbb{C}$.

Exercise 2.21 in § 2.2.2 will give a matrix version of this example. Comparing these two, we are led to the identification of a matrix as a Cartesian coordinate of its entries. Let $\mathbb{R}[X]$ denote the ring of all polynomials with real coefficients, where the polynomial ring is nothing other than the direct sum $\sum \mathbb{R}$ of infinitely many copies of $\mathbb{R}$, which is a subset of the Cartesian product.

Example 1.5. (The third construction) Let j denote a root (in the algebraic closure of $\mathbb{R}$ **provided that it exists**) of the irreducible polynomial

$X^2 + 1$ over $\mathbb{R}$, irreducible because for any real number α, $\alpha^2 + 1 > 0$ and $X^2 + 1$ cannot be decomposed into a product of linear factors. The adjoint $\mathbb{R}(j) = \{a + bj | a, b \in \mathbb{R}\}$ is a field, which is seen to be isomorphic to $\mathbb{C}$.

Example 1.6. (The fourth construction) Let $\mathbb{R}[X]$ denote the ring of all polynomials with real coefficients, where the polynomial ring is nothing other than the direct sum $\sum \mathbb{R}$ (see below).

The factor ring $\mathbb{R}[X]/(X^2 + 1)$ forms a field which is isomorphic to the adjoint $\mathbb{R}(j) = \{a + bj | a, b \in \mathbb{R}\}$ in Example 1.5.

Since

$$\mathbb{R}[X]/(X^2 + 1) = \{a + bX \bmod (X^2 + 1) | a, b \in \mathbb{R}\},$$

we may prove that the mapping $a + bX \to a + bj$ is a field isomorphism.

1.3 Power series and Euler's identity

1.3.1 *Power series*

A **power series** is like a polynomial of infinite degree as given by Remark 1.2 and is of the form

$$\sum_{n=0}^{\infty} a_n z^n,$$

a_n being called the n-th coefficient. They are uniquely determined, i.e. if $\sum_{n=0}^{\infty} a_n z^n = \sum_{n=0}^{\infty} b_n z^n$ in some region, then $a_n = b_n$. Recall that the **geometric series**

$$f(z) = \sum_{n=0}^{\infty} z^n$$

is absolutely and uniformly convergent in $|z| < 1$, divergent in $|z| \geq 1$. The last because on the unit circle, we have $z = e^{2\pi i x}$, $x \in \mathbb{R}$, so that the common ratio is 1 if and only if $x \in \mathbb{Z}$. It turns out that the region of convergence of the power series is always a circle (finite or infinite) and the threshold circle as above is called the **circle of convergence** and its radius r is called the **radius of convergence.** It can be most easily determined by the D'Alembert test:

Theorem 1.6. *The radius of convergence is obtained from*

$$r = \lim_{n \to \infty} \left| \frac{a_n}{a_{n+1}} \right|, \tag{1.11}$$

provided that $a_n \neq 0$.

- Within the circle of convergence, a power series behave exactly like ordinary polynomials, i.e. we may sum, subtract, multiply and divide (provided that the denominator $\neq 0$).
- The multiplication is performed as with polynomials, i.e. one forms the **Cauchy product** $\sum_{m+n=l} a_m b_n$ as the l-th coefficient of the product $(\sum_{m=0}^{\infty} a_m z^m)(\sum_{n=0}^{\infty} b_n z^n)$, which is sometimes referred to as the **Abel convolution** as opposed to the Dirichlet convolution to be introduced below.
- The division of a power series by another one gives rise to a meromorphic function to be studied in §1.3.3 below. The division can also be performed formally as with polynomials.

Example 1.7. The radius of convergence of the power series

$$\sum_{n=0}^{\infty} \frac{z^n}{n!} \tag{1.12}$$

is infinity, i.e. it is absolutely and uniformly convergent over the whole complex plane and defines an analytic function. For $z = x \in \mathbb{R}$, (1.12) coincides with the Taylor expansion for the exponential function e^x, so that (1.12) is the only way of continuing e^x to an analytic function (the principle of analytic continuation). Thus this gives a good motivation to denote (1.12) by e^z (or often $\exp z$) and call it the (complex) **exponential function**:

$$e^z = \sum_{n=0}^{\infty} \frac{z^n}{n!}. \tag{1.13}$$

Similarly, we can define the complex **sine function** and the **cosine function**, respectively by

$$\sin z = \sum_{n=0}^{\infty} \frac{(-1)^n z^{2n+1}}{(2n+1)!}, \quad \cos z = \sum_{n=0}^{\infty} \frac{(-1)^n z^{2n}}{(2n)!}. \tag{1.14}$$

Remark 1.3. To determine the Taylor coefficients we may apply the method of undetermined coefficients which will be again applied to find the residue in (1.41). E.g. supposing the differentiation formula $(e^z)' = e^z$ and the value $e^0 = 1$, we may determine the coefficients a_n for $e^z = \sum_{n=0}^{\infty} a_n z^n$ as follows. First the y-intercept is $a_0 = e^0 = 1$. Differentiating k-times, we get $e^z = \sum_{n=k}^{\infty} n(n-1)\cdots(n-k+1)a_k z^{n-k} = \sum_{n=0}^{\infty} \frac{(n+k)!}{n!} a_{n+k} z^n$, whence $a_n = \frac{1}{n!}$. The same method applies to the sine and cosine functions, but for them, it is far simpler to appeal to Euler's identity below.

Example 1.8. Since the power series (1.13) is absolutely convergent, we may form the Cauchy product for e^{z_1} and e^{z_2} to deduce the **exponential law**:

$$e^{z_1}e^{z_2} = e^{z_1+z_2}. \tag{1.15}$$

We may say that "the binomial theorem implies the exponential law".

From (1.20) and (1.15) we may deduce a conventional definition of the complex exponential function ($z = x + iy \in \mathbb{C}$, $x, y \in \mathbb{R}$)

$$e^z = e^x e^{iy} = e^x(\cos y + i \sin y). \tag{1.16}$$

We note that adopting (1.16) as the definition of the exponential function is problematic because we are left with the same type of problem as to how we define the function e^{iy}, and we are convinced that the above is the only proper way of introducing the exponential function. We can also see that

$$e^{z+2\pi in} = e^z, \; n \in \mathbb{Z}, \tag{1.17}$$

i.e. the exponential function is a periodic function of period $2\pi i$. This periodicity of e^z of period $2\pi i$ is reflected on the multi-valuedness of its inverse function, $\log z$ (cf. §1.7).

Exercise 1.1. Prove (1.15) and in particular the exponential law for

$$e^{i\alpha} \cdot e^{i\beta} = e^{i(\alpha+\beta)}. \tag{1.18}$$

Solution. It suffices to prove (1.15).

$$e^{z_1}e^{z_2} = \sum_{n=0}^{\infty} \frac{1}{n!} \sum_{l+m=n} \frac{n!}{\ell! m!} {z_1}^{\ell} {z_2}^{m} = \sum_{n=0}^{\infty} \frac{1}{n!}(z_1+z_2)^n = e^{z_1+z_2}.$$

1.3.2 *Euler's identity*

In the expansion formula (1.13) for e^z, we substitute ix and classify the terms into 4 classes according to the values of i^n to obtain

$$e^{ix} = \sum_{n=0}^{\infty} \frac{(-1)^n}{(2n)!} x^{2n} + i \sum_{n=0}^{\infty} \frac{(-1)^n}{(2n+1)!} x^{2n+1} = \cos x + i \sin x, \tag{1.19}$$

which is called **Euler's identity**. For $z \in \mathbb{C}$ we obtain the (general) **Euler's identity**:

$$e^{iz} = \cos z + i \sin z. \tag{1.20}$$

In the scope of real analysis, exponential functions and trigonometric functions have nothing to do with one another, but in complex analysis, they are almost the same. Indeed, we have $\cos(-z) = \sum_{n=0}^{\infty} \frac{(-1)^n}{(2n)!}(-z)^{2n} = \cos z$ and $\sin(-z) = -\sin z$, so that $e^{-iz} = \cos z - i \sin z$. Hence

$$\cos z = \frac{1}{2}\left(e^{iz} + e^{-iz}\right), \quad \sin z = \frac{1}{2i}\left(e^{iz} - e^{-iz}\right), \tag{1.21}$$

i.e. $\cos z$, $\sin z$ are expressed in terms of e^{iz}, $e^{-iz} = \frac{1}{e^{iz}}$.

Remark 1.4. To find power series expansions (1.14) for the sine and cosine functions, it is the easiest to use (1.21). It is quite instructive to draw the figure of (1.21) when $z = \theta$ indicates an angle. Indeed, the vector $\boldsymbol{z} = e^{i\theta}$ starts from the origin reaching to the point on the unit circle on the complex plane. The vector $e^{-i\theta}$ is the reflection of $\boldsymbol{z}$ so that their sum divided by 2 is exactly the vector on the real axis pointing to $\cos\theta$. The case of the sine function is more complicated. Express it in the form $\frac{1}{2}\left(e^{i\theta} - e^{-i\theta}\right)e^{-\frac{\pi}{2}i}$. Then we may understand the first difference is the vector on the imaginary axis pointing to $\sin\theta$. Then the factor $e^{-\frac{\pi}{2}i}$ means the clockwise rotation by $\frac{\pi}{2}$ and it amounts to viewing the vector on the imaginary axis on the real axis, thus giving $\sin\theta$.

This clockwise rotation by $\frac{\pi}{2}$ has a more far-reaching effect. Modular functions are defined for the upper half-plane $\mathcal{H} = \{\tau | \operatorname{Im}\tau > 0\}$ and rotating it by $\frac{\pi}{2}$ is given by $s = -i\tau$. Then the variable s lies in the right half-plane

$$\mathcal{RHP} = \{s | \operatorname{Re} s > 0\}. \tag{1.22}$$

Cf. Example 2.12. In this way, the modular functions correspond to zeta-functions which is often referred to as the Hecke correspondence.

Together with (2.124) we have the correspondence

$$\mathcal{RHP} \leftrightarrow \mathcal{H} \leftrightarrow 0 < |q| < 1. \tag{1.23}$$

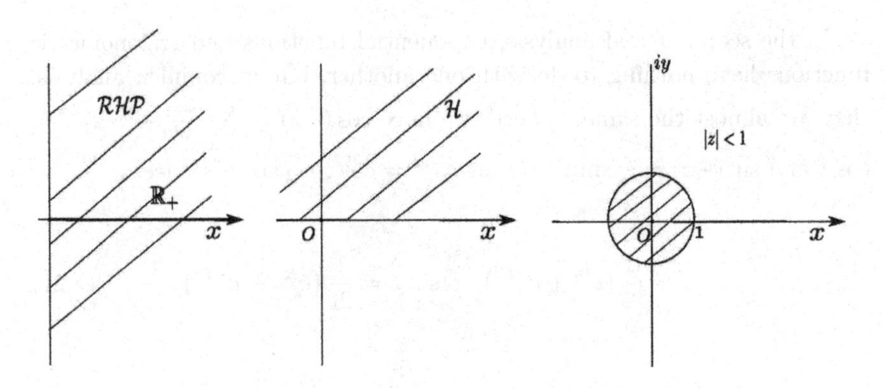

Fig. 1.4 Corresponding domains

In view of (1.21), we may introduce the **hyperbolic functions** by

$$\cosh x = \cos ix = \frac{e^x + e^{-x}}{2}, \quad \sinh x = -i \sin ix = \frac{e^x - e^{-x}}{2}, \tag{1.24}$$
$$\tanh x = \frac{\sinh x}{\cosh x} = \frac{e^x - e^{-x}}{e^x + e^{-x}}.$$

These functions will be used quite often without notice in what follows.

Example 1.9. (A) By (1.21) and the binomial theorem,

$$\cos^3 x = \left(\frac{e^{ix} + e^{-ix}}{2}\right)^3 = \frac{1}{4}\left(\frac{e^{3ix} + e^{-3ix}}{2} + 3\frac{e^{ix} + e^{-ix}}{2}\right),$$

whence $\cos^3 x = \frac{1}{4}(\cos 3x + 3\cos x)$.
(B) Similarly,

$$\sin^3 x = \left(\frac{e^{ix} - e^{-ix}}{2i}\right)^3 = -\frac{1}{4}\left(\frac{e^{3ix} - e^{-3ix}}{2i} - 3\frac{e^{ix} - e^{-ix}}{2i}\right),$$

whence $\sin^3 x = \frac{1}{4}(-\sin 3x + 3\sin x)$. Of course, these can be easily proved by the triplication formulas.

Exercise 1.2. (A) Use (1.21) to deduce the following formulas.
(i) $\cos^4 x = \frac{1}{8}(\cos 4x + 4\cos 2x + 3)$, $\quad \sin^4 x = \frac{1}{8}(\cos 4x - 4\cos 2x + 3)$.
(ii)

$$\cos^5 x = \frac{1}{16}(\cos 5x + 5\cos 3x + 10\cos x),$$
$$\sin^5 x = \frac{1}{16}(\sin 5x - 5\sin 3x + 10\sin x).$$

(iii)

$$\cos^6 x = \frac{1}{32}\left(\cos 6x + 6\cos 4x + 15\cos 2x + 10\right),$$
$$\sin^6 x = -\frac{1}{32}\left(\cos 6x - 6\cos 4x + 15\cos 2x - 10\right).$$

(iv)

$$\cos^7 x = \frac{1}{64}\left(\cos 7x + 7\cos 5x + 21\cos 3x + 35\cos x\right),$$
$$\sin^7 x = -\frac{1}{64}\left(\sin 7x - 7\sin 5x + 21\sin 3x - 35\sin x\right).$$

(v)

$$\cos^8 x = \frac{1}{128}\left(\cos 8x + 8\cos 6x + 28\cos 4x + 56\cos 2x + 35\right),$$
$$\sin^8 x = \frac{1}{128}\left(\cos 8x - 8\cos 6x + 28\cos 4x - 56\cos 2x + 35\right).$$

(B) Evaluate the following definite integrals on the basis of (A). Also evaluate them by recurrences.
(i) $\int_0^{\frac{\pi}{6}} \sin^6 x\,\mathrm{d}x$ (ii) $\int_0^{\frac{\pi}{6}} \cos^6 x\,\mathrm{d}x$ (iii) $\int_0^{\frac{\pi}{4}} \sin^6 x\,\mathrm{d}x$ (iv) $\int_0^{\frac{\pi}{3}} \sin^5 x\,\mathrm{d}x$ (v) $\int_0^{\frac{\pi}{3}} \cos^5 x\,\mathrm{d}x$ (vi) $\int_0^{\frac{\pi}{4}} \sin^5 x\,\mathrm{d}x$ (vii) $\int_0^{\frac{\pi}{2}} \sin^6 x\,\mathrm{d}x$ (viii) $\int_0^{\frac{\pi}{2}} \cos^6 x\,\mathrm{d}x$ (ix) $\int_0^{\frac{\pi}{6}} \cos^4 x\,\mathrm{d}x$ (x) $\int_0^{\frac{\pi}{6}} \sin^4 x\,\mathrm{d}x$ (xi) $\int_0^{\frac{\pi}{4}} \sin^4 x\,\mathrm{d}x$.

Exercise 1.3. Use (1.21) to prove the following equalities.
(i)

$$\cos^{2m} x = \frac{1}{4^m}\left(\binom{2m}{m} + 2\sum_{k=0}^{m-1}\binom{2m}{k}\cos 2\left(m-k\right)x\right)$$

(ii)

$$\sin^{2m} x = \frac{1}{4^m}\left(\binom{2m}{m} + 2\sum_{k=0}^{m-1}(-1)^{m-k}\binom{2m}{k}\cos 2\left(m-k\right)x\right)$$

(iii)

$$\cos^{2m-1} x = \frac{1}{4^{m-1}}\left(\sum_{k=0}^{m-1}\binom{2m-1}{k}\cos\left(2\left(m-k\right)-1\right)x\right)$$

(iv)

$$\sin^{2m-1} x = \frac{1}{4^{m-1}}\left(\sum_{k=0}^{m-1}\binom{2m-1}{k}(-1)^{m-k+1}\sin(2(m-k)-1)x\right).$$

Proposition 1.1. *The addition theorem*

$$\begin{cases} \sin(\alpha+\beta) = \sin\alpha\cos\beta + \cos\alpha\sin\beta, \\ \cos(\alpha+\beta) = \cos\alpha\cos\beta - \sin\alpha\sin\beta \end{cases} \tag{1.25}$$

for trigonometric functions amounts to the exponential law (1.18) *for the complex exponential function (under Euler's identity)*

$$e^{i\alpha} \cdot e^{i\beta} = e^{i(\alpha+\beta)}.$$

Proof. From Euler's identity,

$$\begin{aligned} e^{i\alpha} \cdot e^{i\beta} &= (\cos\alpha + i\sin\alpha)(\cos\beta + i\sin\beta) \\ &= \cos\alpha\cos\beta - \sin\alpha\sin\beta + i(\sin\alpha\cos\beta + \cos\alpha\sin\beta). \end{aligned}$$

By (1.18) this is equal to

$$e^{i(\alpha+\beta)} = \cos(\alpha+\beta) + i\sin(\alpha+\beta)$$

by Euler's identity. Hence comparing the real and imaginary parts, we prove our proposition. □

Another proof. We prove Proposition 1.1 by the consistency theorem 1.2. Consider two functions in z

$$f(z) = \sin(z+u), g(z) = \sin z\cos u + \cos z\sin u, \tag{1.26}$$

for any $u \in \mathbb{R}$. For $z = x \in \mathbb{R}$, we may take $f(x) - g(x) = 0$ for granted by the ordinary addition theorem. Hence by the consistency theorem, $f(z) - g(z) \equiv 0$ on $\mathbb{C}$, i.e.

$$\sin(z+u) = \sin z\cos u + \cos z\sin u \tag{1.27}$$

for all $u \in \mathbb{R}$ and $z \in \mathbb{C}$. Then consider two functions in $w \in \mathbb{C}$

$$f_1(w) = \sin(z+w), g_1(w) = \sin z\cos w + \cos z\sin w \tag{1.28}$$

for any fixed $z \in \mathbb{C}$. By (1.27), $f_1(w) = g_1(w)$ for all $w = u \in \mathbb{R}$, so that $f_1(w) = g_1(w)$ for all $w \in \mathbb{C}$. I.e. the first equality in (1.25) holds true for all $\alpha, \beta \in \mathbb{C}$.

Exercise 1.4. Deduce the exponential law (1.15) by the same reasoning as above.

The polar form in §2.2.2 can be concisely expressed in view of Euler's identity as

$$z = re^{i\theta}, \quad r = |z|, \quad \theta = \arg z. \tag{1.29}$$

Especially, for $n \in \mathbb{Z}$, de Moivre's formula

$$(\cos\theta + i\sin\theta)^n = \cos n\theta + i\sin n\theta \tag{1.30}$$

is nothing but the exponential law

$$(e^{i\theta})^n = e^{in\theta}. \tag{1.31}$$

The de Moivre's formula is extensively used in §2.2.2.

Exercise 1.5. For $\mathbb{R} \ni x \neq 2n\pi$ $(n \in \mathbb{Z})$ prove the formulas

(i)

$$\sum_{k=1}^{n} e^{ikx} = e^{ix}\frac{1-e^{inx}}{1-e^{ix}} = \frac{\sin\left({}^{nx}\!/\!_2\right)}{\sin\left({}^{x}\!/\!_2\right)} e^{i(n+1)x/2}.$$

(ii)

$$\sum_{k=1}^{n} \sin kx = \sin\frac{nx}{2}\sin(n+1)\frac{x}{2}\Big/\sin\frac{x}{2},$$

$$\sum_{k=1}^{n} \cos kx = \sin\frac{nx}{2}\cos(n+1)\frac{x}{2}\Big/\sin\frac{x}{2}.$$

(iii)

$$\sum_{k=1}^{n} e^{i(2k-1)x} = e^{-ix}\sum_{k=1}^{n} e^{i2kx} = \frac{\sin nx}{\sin x}e^{inx}.$$

Solution. (i) The first equality follows from the sum formula (2.2) for a geometric sequence and the second follows by factoring out $e^{\frac{n+2}{2}xi}$ and $e^{\frac{1}{2}xi}$ from the numerator and the denominator, respectively and then applying (1.21).

Exercise 1.6. In the circle $|z| < \frac{\pi}{2}$ we have the Maclaurin expansion

$$\tan z = \sum_{n=0}^{\infty} \frac{A_{2n+1}}{(2n+1)!}z^{2n+1}, \quad |z| < \frac{\pi}{2}. \tag{1.32}$$

The coefficients $\{A_{2n+1}\}$ are called **tangent numbers**.

Verify

$$A_1 = 1, A_3 = 2, A_5 = 16, A_7 = 272, A_9 = 7936, A_{11} = 353792, \cdots \tag{1.33}$$

in three different ways. Hence (1.32) reads

$$\tan x = x + \frac{1}{3}x^3 + \frac{2}{15}z^5 + \frac{17}{315}z^7 + \cdots. \tag{1.34}$$

Solution. The most naive method would be to use the expansions for sin and cos to divide the former by the latter, or writing cos tan = sin and comparing the coefficients. This method may lead to (1.34) with much effort but may not give much insight into the general structure of A_{2n+1}.

The second method uses the equality

$$f' = 1 + f^2, \tag{1.35}$$

which is equivalent to the Pythagoras theorem $1 + \tan^2\theta = \frac{1}{\cos^2\theta}$. Hence $f(0) = 0, f'(0) = 1$. In computing higher derivatives, we substitute (1.35) whenever we encounter f'. We obtain successively

$$\begin{aligned}
&f'' = 2f + 2f^3,\ f^{(3)} = 2 + (2+3!)f^2 + 3!f^4,\\
&f^{(4)} = 2(2+3!)f + (2(2+3!) + 4!))f^3 + 4!f^5,\\
&f^{(5)} = 2(2+3!)\\
&+ (2(2+3!) + 3(2(2+3!) + 4!)f^2 + (3(2(2+3!) + 4!) + 5!)f^3 + 5!f^6,\\
&f^{(6)} = 2(2(2!+3!) + 3(2(2!+3!) + 4!))f(1+f^2) + \cdots,\\
&f^{(7)} = 2(2(2+3!) + 3(2(2+3!) + 4!)) + \cdots,
\end{aligned}$$

which verifies (1.34). This method gives a nest structure of the coefficients, but it's not easy to go on in this fashion.

The third method relies on Euler numbers to be introduced below.

1.3.3 *Laurent expansion, residues*

Cf. §2.10.

Definition 1.2. Suppose a function $f(z)$ has a denominator $(z-\alpha)^k$, $k \in \mathbb{N}$ at $z = \alpha$ such that $f(z) = \frac{f_1(z)}{(z-\alpha)^k}$ and $f_1(z)$ is analytic at α and $\neq 0$. In this case, we say f has a **pole** of order k at $z = \alpha$. If $k = 1$, the pole is called **simple**. Functions with only poles as their singularities are called **meromorphic functions.**

Cf. Definition 1.3 for a meromorphic function with finitely many poles. Since $f_1(z) = \sum\limits_{n=0}^{\infty} a_n(z-\alpha)^n$, we have

$$f(z) = \frac{a_0}{(z-\alpha)^k} + \cdots + \frac{a_{k-1}}{z-\alpha} + \sum_{n=k}^{\infty} a_n(z-\alpha)^{n-k},$$

or

$$f(z) = \frac{a_{-k}}{(z-\alpha)^k} + \cdots + \frac{a_{-1}}{z-\alpha} + h(z), \tag{1.36}$$

with slight change of notation $a_j \to a_{j-k}$, where $h(z)$ is analytic at $z = \alpha$ and is often referred to as a **holomorphic part**, while the singular part consisting of fractions is called the **principal part**.

Example 1.10. The function

$$f(z) = \frac{1}{e^z - 1}$$

has a simple pole at $z = 0$, the Laurent series being

$$\frac{1}{e^z - 1} = \frac{1}{z + \frac{z^2}{2!} + \frac{z^3}{3!} + \cdots} = \frac{1}{z} + f_0(z), \tag{1.37}$$

where $f_0(z) = -\frac{1}{2} + \frac{1}{12}z + \cdots$ is the holomorphic part (at $z = 0$). Since $e^z = 1$ occurs for $z = 2\pi in$, $n \in \mathbb{Z}$, the Laurent expansion (1.37) holds in the annulus $0 < |z| < 2\pi$.

Suppose $f(z)$ is analytic except for $z = \alpha$ in a domain and let C be a circle inside D with center at α. When we evaluate the integral $\int_C f(z)\,\mathrm{d}z$ along C, we apply (1.8) to get

$$\int_C f(z)\,\mathrm{d}z = 2\pi i a_{-1}, \tag{1.38}$$

and a_{-1} is the most important quantity called the **residue** of $f(z)$ at $z = \alpha$:

$$a_{-1} = \operatorname*{Res}_{z=\alpha} f(z). \tag{1.39}$$

The Cauchy residue theorem, Theorem 1.7, asserts that (1.38) is true for any piecewise smooth curve encircling the point α in which $f(z)$ is analytic except at $z = \alpha$.

To find the residue is rather simple. We use the **method of undetermined coefficients**. Clearing the denominators of the right-hand side of (1.36), by multiplying by $(z - \alpha)^k$, we get

$$(z - \alpha)^k f(z) = a_{-k} + \cdots + a_{-1}(z - \alpha)^{k-1} + h_1(z), \tag{1.40}$$

$h_1(z)$ being holomorphic and $h_1(\alpha) = \cdots = h_1^{(k-1)}(\alpha) = 0$.

Differentiating (1.40) $(k - 1)$-times, we obtain

$$\frac{1}{(k-1)!}\frac{\mathrm{d}^{k-1}}{\mathrm{d}z^{k-1}}\left\{(z - \alpha)^k f(z)\right\} = a_{-1} + h_2(z),$$

whence,

$$\operatorname*{Res}_{z=\alpha} f(z) = \lim_{z \to \alpha} \frac{1}{(k-1)!}\frac{\mathrm{d}^{k-1}}{\mathrm{d}z^{k-1}}\left\{(z - \alpha)^k f(z)\right\}. \tag{1.41}$$

Combining (1.39) and (1.41), we obtain

$$\int_C f(z)\mathrm{d}z = 2\pi i \operatorname*{Res}_{z=\alpha} f(z) = \frac{2\pi i}{(k-1)!} \lim_{z\to\alpha} \frac{\mathrm{d}^{k-1}}{\mathrm{d}z^{k-1}} \left\{(z-\alpha)^k f(z)\right\}, \quad (1.42)$$

i.e. we may conduct "integration" by "differentiating."

E.g.,

$$\operatorname*{Res}_{z=i} \frac{1}{z^2+1} = \lim_{z\to i} (z-i) \frac{1}{(z-i)(z+i)} = -\frac{i}{2}.$$

Example 1.11. We integrate the function

$$f(z) = \frac{z}{e^z - 1}, \quad (1.43)$$

along the circle C with center at $2\pi i$ and radius π. Then

$$\begin{aligned}\int_C f(z)\,\mathrm{d}z &= 2\pi i \operatorname*{Res}_{z=2\pi i} f(z) = 2\pi i \lim_{z\to 2\pi i} \frac{z-2\pi i}{e^z - e^{2\pi i}} z \\ &= 2\pi i \frac{1}{(e^z)'\,|_{z=2\pi i}} 2\pi i = -4\pi^2.\end{aligned}$$

1.4 Residue calculus

We are in a position to state the Cauchy residue theorem, one of the fundamental theorems in complex analysis.

Theorem 1.7. (Cauchy residue theorem) *Let C be a piecewise smooth closed Jordan curve whose interior is D and let $f(z)$ be analytic in D except for a finite number of poles at α_k, $1 \le k \le n$. Then*

$$\int_C f(z)\mathrm{d}z = 2\pi i \sum_{k=1}^{n} \operatorname*{Res}_{z=\alpha_k} f(z).$$

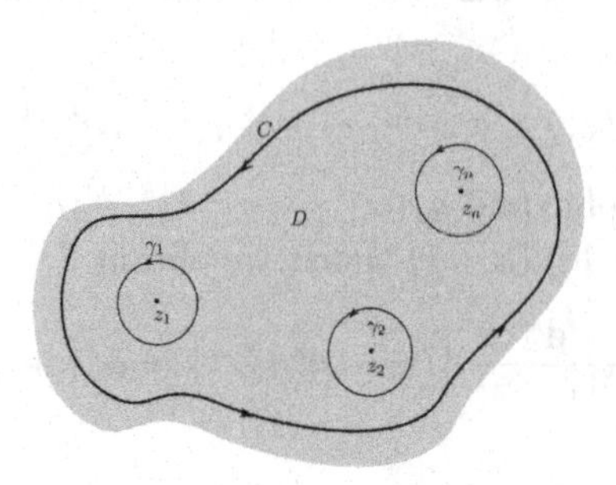

Fig. 1.5 The domain D with finitely many poles

The proof is similar to the one we gave in the contraction principle after Example 1.1. We encircle all the poles α_k by small circles γ_k with center at α_k. Then we connect the circle γ_k to γ_{k+1} and γ_1 and γ_{k+1} to C) at two points by mutually non-intersecting polygons (piece-wise smooth arcs) and then cut the interior into two. Then the integral along C is the same as the sum of integrals along those small circles. Then (1.8) allows us to evaluate these integrals.

Example 1.12. (Lagrange interpolation formula) Given $n+1$ different numbers α_k and points $(\alpha_k, \beta_k) \in \mathbb{C}^2$, $0 \le k \le n$, we find the polynomial $f(X)$ of degree n which passes through these points. Let

$$g(X) = \prod_{k=0}^{n} (X - \alpha_k) \tag{1.44}$$

which is the monic polynomial that has zeros at α_k's.

Consider the fraction $\frac{f(X)}{g(X)}$, where $f \in \mathbb{C}[X]$ of degree n. The rational function $\frac{f(z)}{g(z)}$, having a (at most) simple pole at α_k's, is of the form $\sum_{k=0}^{n} \frac{a_k}{z-\alpha_k}$:

$$\frac{f(z)}{g(z)} = \sum_{k=0}^{n} \frac{a_k}{z - \alpha_k}. \tag{1.45}$$

Hence

$$\begin{aligned} a_k &= \operatorname*{Res}_{z=\alpha_k} \frac{f(z)}{g(z)} = \lim_{z \to \alpha_k} \frac{f(z)}{\prod_{j \ne k}(z - \alpha_j)} = \frac{f(\alpha_k)}{\prod_{j \ne k}(\alpha_k - \alpha_j)} \\ &= \frac{f(\alpha_k)}{g'(\alpha_k)}. \end{aligned} \tag{1.46}$$

Hence if we choose

$$a_k = \frac{\beta_k}{g'(\alpha_k)},$$

then the polynomial

$$f(X) = \sum_{k=0}^{n} \frac{g(X)}{X - \alpha_k} \frac{\beta_k}{g'(\alpha_k)}, \tag{1.47}$$

has the value β_k at $z = \alpha_k$. (1.47) is known as the **Lagrange interpolation formula.**

Remark 1.5. In some books, (1.47) is proved as follows. First considering the polynomial

$$f(X) = \sum_{k=0}^{n} a_k \prod_{j \neq k} (X - \alpha_j) \tag{1.48}$$

and one determines the coefficient a_k by setting $X = \alpha_k$. However, (1.48) is precisely (1.45) with cleared denominator and if one just goes through this easy proof, one would lose the essence of the residue calculus method underlying the Lagrange method of interpolation.

Proposition 1.2. *Let $L = K(\alpha)$ be a finite separable extension of degree n and let $f(X) = \mathrm{Irr}(\alpha, K, X)$ be the irreducible polynomial of α over K. Let f' be the derivative of f and*

$$\frac{f(X)}{X - \alpha} = \sum_{k=0}^{n-1} b_k X^k. \tag{1.49}$$

Then the dual basis of $1, \alpha, \ldots, \alpha^{n-1}$ is

$$\frac{b_k}{f'(\alpha)}, \qquad 0 \leq k \leq n-1, \tag{1.50}$$

where the dual basis means that

$$\mathrm{Tr}\left(\alpha^i \frac{b_j}{f'(\alpha)}\right) = \delta_{ij} \tag{1.51}$$

and where the trace means the sum of all conjugates.

Proof. Let $\alpha_1, \ldots, \alpha_n$ be the distinct roots of f. Then by Example 1.12, we have

$$\sum_{k=1}^{n} \frac{f(X)}{X - \alpha_k} \frac{\alpha_k^r}{f'(\alpha_k)} = X^r, \quad 0 \leq r \leq n-1. \tag{1.52}$$

Defining the trace of a polynomial with coefficients in L to the polynomial obtained by applying the trace to the coefficients, we deduce from (1.52) that

$$\mathrm{Tr}\left(\frac{f(X)}{X - \alpha} \frac{\alpha^i}{f'(\alpha)}\right) = X^i, \quad 0 \leq i \leq n-1. \tag{1.53}$$

Comparing the coefficients of each power of X in (1.53), we deduce (1.50), completing the proof. □

The following lemma is a special case of Lemma 1.2 below. In the beginning we advise you to use this version since you will not forget to count the residue in this way.

Lemma 1.1. *If $f(z)$ has a simple pole at $z = a \in \mathbb{R}$ and γ_a is a lower semi-circle in the positive direction with a small radius r tending to 0, then*

$$\int_{\gamma_a} f(z)\,\mathrm{d}z \to 2\pi i \frac{1}{2} \operatorname*{Res}_{z=a} f(z) \tag{1.54}$$

as $r \to +0$.

Proof. The lower semi-circle has the representation

$$\gamma_a : z - a = re^{it}, \quad \pi \le t \le 2\pi$$

and near $z = a$, $f(z)$ is of the form

$$f(z) = \frac{a_{-1}}{z-a} + g(z), \tag{1.55}$$

where $g(z)$ is holomorphic at $z = a$ and $a_{-1} = \operatorname{Res}_{z=a} f(z)$.

Hence

$$\int_{\gamma_a} f(z)\,\mathrm{d}z = \int_{\pi}^{2\pi} f(a + re^{it}) ire^{it}\,\mathrm{d}t = i \int_{\pi}^{2\pi} \left(a_{-1} + g(a + re^{it})re^{it}\right) \mathrm{d}t \tag{1.56}$$

$$\to i\pi a_{-1} = 2\pi i \frac{1}{2} \operatorname*{Res}_{z=a} f(z)$$

as $r \to 0+$, completing the proof. □

Example 1.13. We find the value of the integral

$$\int_0^{\infty} \frac{\sin x}{x}\,\mathrm{d}x = \frac{\pi}{2}. \tag{1.57}$$

Cf. Example 1.30 for another solution. We integrate the function $f(z) = \frac{e^{iz}}{z}$ around the contour C consisting of the upper semi-circle C_R of radius R going to infinity, the line segment $[-R, -r]$, the lower semi-circle γ_0 of radius r going to 0 and the line segment $[r, R]$. then by the residue theorem, we have

$$\int_C f(z)\,\mathrm{d}z = 2\pi i \operatorname*{Res}_{z=0} \frac{e^{iz}}{z} = 2\pi i. \tag{1.58}$$

On the other hand,

$$\int_C f(z)\,\mathrm{d}z = \int_{C_R} f(z)\,\mathrm{d}z + \int_{-R}^{-r} f(x)\,\mathrm{d}x + \int_r^R f(x)\,\mathrm{d}x \tag{1.59}$$

$$+ \int_{\gamma_a} f(z)\,\mathrm{d}z \to 2\pi i \frac{1}{2} \operatorname*{Res}_{z=0} f(z) + \mathrm{PV} \int_{-\infty}^{\infty} f(x)\,\mathrm{d}x$$

as $R \to \infty$, $r \to 0+$, where PV $\int$ indicates the Cauchy principal value to be introduced in (1.217).

By (1.58) and (1.59), we conclude (1.57).

The following examples illustrate Theorem 1.5, i.e. expandability inside the biggest circle contained in the domain of analyticity.

Example 1.14. We note that the function in (1.43) is analytic in $|z| < 2\pi$ if we modify the value of f at $z = 0$ to be 1, thus removing the removable singularity at $z = 0$. The nearest singularities of f from the origin are $z = \pm 2\pi i$, and so the domain of analyticity is $|z| < 2\pi$. Its Taylor coefficients give rise to an important rational sequence $\{B_n\}$ called **Bernoulli numbers**:

$$\frac{z}{e^z - 1} = \sum_{n=0}^{\infty} \frac{B_n}{n!} z^n, \quad |z| < 2\pi. \tag{1.60}$$

In view of this, we call $\frac{z}{e^z-1}$ the **generating function (generating power series)** of Bernoulli numbers.

- One way of describing analytic number theory is finding the relation between generating Dirichlet series.
- **Generating Dirichlet series** are mostly zeta-functions which have zeta symmetry in conjunction with some gamma factors, and so the special functions come in the picture at this stage.

Exercise 1.7. Give details of the reasoning in Example 1.11.

Solution. Since near $z = 0$, $f(z) = 1 - \frac{1}{2}z + \frac{1}{12}z^2 + \cdots$ by Example 1.10, the function $f(z) = \frac{z}{e^z-1}$ has a removable singularity at $z = 0$. Hence the modified function

$$g(x) = \begin{cases} 1, & z = 0 \\ \dfrac{z}{e^z - 1}, & z \neq 0 \end{cases}$$

is analytic inside the circle $|z| < 2\pi$.

Example 1.15. The product $\frac{z}{e^z-1}e^{xz}$ is analytic in $|z| < 2\pi$ and has a Taylor expansion of the form:

$$\frac{ze^{xz}}{e^z - 1} = \sum_{n=0}^{\infty} \frac{B_n(x)}{n!} z^n, \ |z| < 2\pi. \tag{1.61}$$

The coefficient $B_n(x)$ is called the (n-th) **Bernoulli polynomial**. Thus the function $\frac{ze^{xz}}{e^z-1}$ is the generating power series for them. In the same way, we may define the **Euler polynomials** $E_n(x)$ by

$$\frac{2e^{xz}}{e^z+1} = \sum_{n=0}^{\infty} \frac{E_n(x)}{n!} z^n, \ |z| < \pi \tag{1.62}$$

the series being absolutely and uniformly convergent in $|z| < \pi$ (the singularity from the origin being $z = \pm i\pi$).

From (1.61) it follows that the **Bernoulli numbers**

$$B_n = B_n(0). \tag{1.63}$$

Example 1.16. Along with the Bernoulli numbers, we introduce **Euler numbers** $\{E_{2n}\}$ either by

$$\frac{1}{\cosh z} = \frac{2}{e^z + e^{-z}} = \sum_{n=0}^{\infty} \frac{E_{2n}}{(2n)!} z^{2n}, \quad |z| < \frac{\pi}{2} \tag{1.64}$$

or by

$$\sec z = \frac{2}{e^{iz} + e^{-iz}} = \sum_{n=0}^{\infty} \frac{(-1)^n E_{2n}}{(2n)!} z^{2n}, \quad |z| < \frac{\pi}{2}. \tag{1.65}$$

These two are equivalent and shift to each other under the rotation of the complex plane by $\frac{\pi}{2}$. One can easily evaluate the following:

$$E_0 = 1, E_2 = -1, E_4 = 5, E_6 = -61, E_8 = 1385, E_{10} = -50521 \cdots. \tag{1.66}$$

Note that unlike Bernoulli numbers, Euler numbers are all integers. Similarly to (1.63), we have

$$E_n = 2^n E_n \left(\frac{1}{2} \right) \tag{1.67}$$

which follows from (1.62) and (1.64) on noting that $\frac{2}{e^z+e^{-z}} = \frac{2e^z}{e^{2z}+1}$.

On writing

$$(\sec z)^2 = \sum_{n=0}^{\infty} \frac{(-1)^n E_{2n}^{(2)}}{(2n)!} z^{2n}, \quad |z| < \frac{\pi}{2}, \tag{1.68}$$

$E_{2n}^{(2)}$ being called the $2n$-th Euler number of order 2, it follows from (1.65) that

$$E_{2n}^{(2)} = (2n)! \sum_{v_1+v_2=n} \frac{E_{2v_1} E_{2v_2}}{(2v_1)!(2v_2)!}. \tag{1.69}$$

Comparing (1.32) to (1.68), we find that

$$E_{2n}^{(2)} = (-1)^n A_{2n+1}. \tag{1.70}$$

Using (1.69), (1.70), we see immediately

$$A_5 = 4!\left(2 \cdot \frac{E_0 E_4}{4!} + \frac{E_2 E_2}{2!2!}\right) = 16,$$

$$A_7 = -2 \cdot 6!\left(\frac{E_0 E_6}{6!} + \frac{E_2 E_4}{2!4!}\right) = 272,$$

so that

$$\frac{A_5}{5!} = \frac{2}{15}, \quad \frac{A_7}{7!} = \frac{17}{315}, \dots.$$

Example 1.17. Suppose there is an object with the closed boundary curve C in the flow consisting of three components

- The parallel flow $w = Uz$, $U > 0$
- The circulation $w = -\frac{\Gamma}{2\pi i}\log z$, $\Gamma > 0$
- The flow caused by the object $w = \sum_{n=1}^{\infty} \frac{c_n}{z^n}$

By the principle of superposition, the flow around the object may be expressed as

$$w = Uz - \frac{\Gamma}{2\pi i}\log z + \sum_{n=1}^{\infty} \frac{c_n}{z^n}. \tag{1.71}$$

The complex velocity is

$$\frac{\mathrm{d}w}{\mathrm{d}z} = U - \frac{\Gamma}{2\pi i z} - \sum_{n=1}^{\infty} \frac{n c_n}{z^{n+1}}. \tag{1.72}$$

Let X, Y be the components of the force exerting on the object, i.e. X = drag and Y = lift, then we have the **Blasius formula** (1910)

$$X - iY = -\frac{\rho}{2i}\int_C \left(\frac{\mathrm{d}w}{\mathrm{d}z}\right)^2 \mathrm{d}z, \tag{1.73}$$

where ρ indicates the density of the flow.

Since the residue of $\left(\frac{\mathrm{d}w}{\mathrm{d}z}\right)^2$ is clearly $-2\frac{U\Gamma}{2\pi i}$, it follows from the Cauchy residue theorem that

$$X - iY = -i\rho U\Gamma, \tag{1.74}$$

whence we deduce

Theorem 1.8. **(Kutta-Joukovski theorem)** *An object with a sufficiently smooth boundary in the flow given by* (1.71) *receives no* **drag** $X = 0$ *but receives the* **lift** $Y = \rho U\Gamma$.

By Example 1.25, the circulation flow may be produce by rotating a cylinder, whence we have

Corollary 1.1. (**Magnus effect**) *A rotating cylinder with intensity Γ in the flow of uniform speed U with density ρ receives the lift of intensity $\rho U\Gamma$.*

1.4.1 *Cauchy integral formula and its consequences*

The following theorem is also fundamental in complex analysis and has many important consequences, one of which will be utilized in §1.8.

Theorem 1.9. (Cauchy integral formula) *Let $f(z)$ be analytic in a domain with boundary C. Then*

$$f(z) = \frac{1}{2\pi i}\int_C \frac{f(w)}{w-z}\mathrm{d}w, \tag{1.75}$$

i.e. the values of $f(z)$ inside C are determinant by those on C.

Remark 1.6. The reason why we state the Cauchy integral formula here is that it may be viewed as a consequence of the Cauchy residue theorem Theorem 1.7: $\frac{1}{2\pi i}\int_C \frac{f(w)}{w-z}\mathrm{d}w = \mathrm{Res}_{w=z}\frac{f(w)}{w-z} = f(z)$. For that matter, the Cauchy integral theorem, Theorem 1.4 itself may be thought of as a special case of the residue theorem by loosening the meromorphic condition in Definition 1.2 to include the case without the denominator $k=0$ or $a_1=0$ case. Then (1.38) amounts to the Cauchy integral theorem.

Definition 1.3. A function analytic over the whole plane is called an integral (or entire) function. Therefore, a meromophic function $f(z)$ with finitely many poles may be expressed in the form $\frac{h(z)}{P(z)}$, where h is an integral function and P is a polynomial which has zeros at those poles of f.

Many of the familiar functions including exponential functions, and trigonometric functions are all integral functions. When a function has a denominator, we encounter meromorphic functions. Cf. §1.13 for integral functions of exponential type.

Corollary 1.2. (The Cauchy estimate of coefficients) *If $f(z)$ is analytic in $D: |z-z_0|<r$ and $f(z)=O(1)$ on D, then*

$$f^{(n)}(z) = O\left(\frac{n!}{r^n}\right). \tag{1.76}$$

Corollary 1.3. (Liouville's theorem) *A bounded integral function is a constant.*

Proof. We have $f(z) = O(1)$ on $\mathbb{C}$. Hence for any $R > 0$, we have by Corollary 1.2

$$f'(z) = O\left(\frac{1}{R}\right),$$

whence we have $f'(z) = 0$. Hence the Newton-Leibnitz rule (Corollary 1.5) implies that f is a constant. □

Liouville's theorem is often used in determining the expansions of meromorphic functions. E.g. to deduce the partial fraction expansion (1.226) rigorously, we appeal to it. The difference is bounded and integral, so that it must be a constant. But the difference goes to 0 as the variable $s \to \infty$, the constant must be 0. Similar argument may be applied to the deduction of the partial fraction expansions in §1.11.3. As long as we can locate all the poles of a meromorphic function, we may apply Liouville's theorem to determine the partial fraction expansion.

Corollary 1.4. (Fundamental theorem of algebra) *Any polynomial* $f(X) = a_nX^n + \cdots + a_1X + a_0 \in \mathbb{C}[X]$ $(a_n \neq 0)$ *of degree* $n \geq 1$ *has at least one root (zero) in* $\mathbb{C}$*, whence it has roots whose number coincides with* n *up to multiplicity.*

1.5 Review on vector-valued functions

1.5.1 *Differentiation*

Let a vector-valued function $\boldsymbol{y} = \boldsymbol{y}(\boldsymbol{x})$ in a vector argument $\boldsymbol{x} = \begin{pmatrix} x_1 \\ x_2 \\ \vdots \\ x_n \end{pmatrix} \in D \subset \mathbb{R}^n$ be given by

$$\begin{pmatrix} y_1 \\ y_2 \\ \vdots \\ y_m \end{pmatrix} = \boldsymbol{y} = \boldsymbol{y}(\boldsymbol{x}) = \begin{pmatrix} y_1(\boldsymbol{x}) \\ y_2(\boldsymbol{x}) \\ \vdots \\ y_m(\boldsymbol{x}) \end{pmatrix} = \begin{pmatrix} y_1(x_1, x_2, \cdots, x_n) \\ y_2(x_1, x_2, \cdots, x_n) \\ \cdots \\ y_m(x_1, x_2, \cdots, x_n) \end{pmatrix}, \tag{1.77}$$

which is equivalent to the system of equations

$$\begin{cases} y_1 = y_1(\boldsymbol{x}) = y_1(x_1, x_2, \cdots, x_n) \\ y_2 = y_2(\boldsymbol{x}) = y_2(x_1, x_2, \cdots, x_n) \\ \cdots \\ y_m = y_m(\boldsymbol{x}) = y_m(x_1, x_2, \cdots, x_n). \end{cases} \tag{1.78}$$

Definition 1.4. The vector-valued function $\boldsymbol{y} = \boldsymbol{y}(\boldsymbol{x})$ is said to be totally differentiable (or Fréchet differentiable) at $\boldsymbol{x}$ if

$$\boldsymbol{y}(\boldsymbol{x}+\boldsymbol{h}) = \boldsymbol{y}(\boldsymbol{x}) + A\boldsymbol{h} + o(|\boldsymbol{h}|) \tag{1.79}$$

as $\boldsymbol{h} \to \boldsymbol{o}$, i.e. $|\boldsymbol{h}| \to 0$. Here A is an (m,n)-matrix and is called the **gradient** (or the Jacobi matrix) of $\boldsymbol{y}$, denoted $\nabla \boldsymbol{y}$.

$$\nabla \boldsymbol{y} = \nabla \begin{pmatrix} y_1 \\ y_2 \\ \vdots \\ y_m \end{pmatrix} = \begin{pmatrix} \nabla y_1 \\ \nabla y_2 \\ \vdots \\ \nabla y_m \end{pmatrix} = \begin{pmatrix} \frac{\partial y_1}{\partial x_1} & \frac{\partial y_1}{\partial x_2} & \cdots & \frac{\partial y_1}{\partial x_n} \\ \frac{\partial y_2}{\partial x_1} & \frac{\partial y_2}{\partial x_2} & \cdots & \frac{\partial y_2}{\partial x_n} \\ \cdots \\ \frac{\partial y_m}{\partial x_1} & \frac{\partial y_m}{\partial x_2} & \cdots & \frac{\partial y_m}{\partial x_n} \end{pmatrix}. \tag{1.80}$$

In the case of a scalar function $\boldsymbol{y} = y = y(x_1, \ldots, x_n)$, the increment term $A\boldsymbol{h}$ in (1.4) amounts to

$$\mathrm{d}y = \mathrm{d}\boldsymbol{y} = \nabla y \mathrm{d}\boldsymbol{x} = \frac{\partial y}{\partial x_1}\mathrm{d}x_1 + \cdots + \frac{\partial y}{\partial x_n}\mathrm{d}x_n, \tag{1.81}$$

where $\mathrm{d}\boldsymbol{y}$ means the differential of $\boldsymbol{y}$, on writing $\boldsymbol{h} = \mathrm{d}\boldsymbol{x} = (\mathrm{d}x_1, \ldots, \mathrm{d}x_n)$.

$$\mathrm{d}\boldsymbol{y} = \mathrm{d}\boldsymbol{y}(\boldsymbol{x}) = \begin{pmatrix} \mathrm{d}y_1 \\ \mathrm{d}y_2 \\ \vdots \\ \mathrm{d}y_m \end{pmatrix} \tag{1.82}$$

(1.81) is a differential form of degree 1 in n variables.

For more details on the 2-dimensional case cf. §2.5. We note that total differentiability is usually assured by the (partial) differentiability of entries and we usually assume that all the entries are of C^2 class.

1.5.2 *Fluid mechanics*

Cf. Example 1.21 below. Consider a 2-dimensional flow with velocity vector $\boldsymbol{v} = \begin{pmatrix} P \\ Q \end{pmatrix} \in C(D)$, where $D \subset \mathbb{R}^2$ is a domain with its boundary as a closed Jordan curve C. Then the divergence and the curl are defined respectively by

$$\operatorname{div} \boldsymbol{v} = \frac{\partial P}{\partial x} + \frac{\partial Q}{\partial y} \tag{1.83}$$

and

$$\operatorname{curl} \boldsymbol{v} = \omega = \frac{\partial Q}{\partial x} - \frac{\partial P}{\partial y}. \tag{1.84}$$

The 2-dimensional flow occurs when the fluid flows through a long cylindrical tube whose axis is perpendicular to the direction of the flow.

We consider the function

$$W := \operatorname{curl} \boldsymbol{v} + i \operatorname{div} \boldsymbol{v} = \frac{\partial Q}{\partial x} - \frac{\partial P}{\partial y} + i \left(\frac{\partial P}{\partial x} + \frac{\partial Q}{\partial y} \right), \tag{1.85}$$

for which another expression is given in (1.112). We integrate it over D (with $\mathrm{d}A$ denoting the area element):

$$\begin{aligned} \iint_D W \,\mathrm{d}A &= \iint_D \left(\frac{\partial Q}{\partial x} - \frac{\partial P}{\partial y} \right) \mathrm{d}A + i \iint_D \left(\frac{\partial P}{\partial x} + \frac{\partial Q}{\partial y} \right) \mathrm{d}A \qquad (1.86) \\ &= \int_C P\mathrm{d}x + Q\mathrm{d}y + i \int_C -Q\mathrm{d}x + P\mathrm{d}y \end{aligned}$$

by Green's theorem, Theorem 2.12.

We interpret the last two curvilinear integrals of (1.86) as a contour integral in the variable $z = x + iy$, and *a fortiori* $\mathrm{d}z = \mathrm{d}x + i\mathrm{d}y$. Hence

$$\bar{v}\, \mathrm{d}z = (P - iQ)(\mathrm{d}x + i\mathrm{d}y) = P\mathrm{d}x + Q\mathrm{d}y + i(-Q\mathrm{d}x + P\mathrm{d}y). \tag{1.87}$$

I.e.

$$\iint_D W \,\mathrm{d}A = \int_C \bar{v}\, \mathrm{d}z. \tag{1.88}$$

Since

$$v_t = P\frac{\mathrm{d}x}{\mathrm{d}s} + Q\frac{\mathrm{d}y}{\mathrm{d}s}, \quad v_n = -Q\frac{\mathrm{d}x}{\mathrm{d}s} + P\frac{\mathrm{d}y}{\mathrm{d}s}, \tag{1.89}$$

where $v_t = \boldsymbol{v} \cdot \boldsymbol{t}$ is the component in the direction of the tangent vector and v_n the component in the direction of the normal vector, we also have

$$\bar{v}\mathrm{d}z = (v_t + iv_n)\mathrm{d}s. \tag{1.90}$$

From (1.85), (1.88) and (1.90) we conclude that

$$\iint_D (\operatorname{curl} \boldsymbol{v} + i \operatorname{div} \boldsymbol{v}) \, \mathrm{d}A = \int_C (v_{\boldsymbol{t}} + i v_{\boldsymbol{n}}) \, \mathrm{d}s. \tag{1.91}$$

Comparing the real and imaginary parts, we conclude

Theorem 1.10. *If $D \subset \mathbb{R}^2$ is a domain with its boundary curve $\partial D = C$, then*

$$Q = Q(C) = \int_C v_{\boldsymbol{n}} = \iint_D \operatorname{div} \boldsymbol{v}. \tag{1.92}$$

$$\Gamma = \Gamma(C) = \int_C v_{\boldsymbol{t}} = \iint_D \operatorname{curl} \boldsymbol{v}. \tag{1.93}$$

We now assume that the flow is **incompressible** $\operatorname{div} \boldsymbol{v} = 0$ and **irrotational** $\operatorname{curl} \boldsymbol{v} = 0$, respectively. Then by Theorem 1.10, we have $\int_C \bar{v} \, \mathrm{d}z = 0$ for any piecewise smooth Jordan curve C in D. Hence by Morera's theorem 2.17, $\bar{v}(z)$ is analytic in D. Then by Theorem 2.15, there exists a primitive function $\bar{v}(z)$, which is called the **complex potential** of $\bar{v}$.

A real-valued function ϕ such that $\boldsymbol{f} = \nabla \phi$ is called a potential function of $\boldsymbol{f}$. Thus the preceding argument gives rise to a potential function for the velocity $\bar{\boldsymbol{v}} = \begin{pmatrix} P \\ -Q \end{pmatrix}$ corresponding to $\bar{v}$ in (1.110) below.

1.5.3 *Thermodynamic intermission*

Notation
T = absolute temperature
S = entropy to be defined below
p = pressure
V = volume $\left(\frac{\partial S}{\partial T}\right)_V$ = the partial derivative of S as a function in T, V with respect to T with V viewed as a constant in T.

Example 1.18. The state of an object in thermal equilibrium can be described by the variables T, S, p, V. We have the relation

$$\left(\frac{\partial S}{\partial T}\right)_V \left(\frac{\partial p}{\partial V}\right)_S = \left(\frac{\partial S}{\partial T}\right)_p \left(\frac{\partial p}{\partial V}\right)_T. \tag{1.94}$$

This can be easily proved if we view the partial derivative as a Jacobian, e.g.

$$\left(\frac{\partial T}{\partial p}\right)_V = \frac{\partial(T, V)}{\partial(p, V)} = \frac{\partial(V, T)}{\partial(V, p)} = -\frac{\partial(V, T)}{\partial(p, V)}. \tag{1.95}$$

By (1.95), the left-hand side of (1.94) is

$$-\frac{\partial(S,V)}{\partial(T,V)}\frac{\partial(p,S)}{\partial(S,V)},$$

which is $-\frac{\partial(p,S)}{\partial(T,V)}$ by the multiplication theorem for the Jacobians. This last expression may be written as

$$-\frac{\partial(S,p)}{\partial(V,T)} = \frac{\partial(S,p)}{\partial(T,p)}\frac{\partial(p,T)}{\partial(V,T)},$$

which is the right-hand side of (1.94).

Definition 1.5. The interior energy $U = U(S,V)$ of a closed thermodynamical system is a state quantity if and only if it can be expressed as a total differential

$$\mathrm{d}U = \left(\frac{\partial U}{\partial S}\right)_V \mathrm{d}S + \left(\frac{\partial U}{\partial V}\right)_S \mathrm{d}V. \tag{1.96}$$

Example 1.19. The fundamental equation in thermodynamics is

$$\mathrm{d}U = T\mathrm{d}S - p\,\mathrm{d}V \tag{1.97}$$

and the Maxwell equation

$$\left(\frac{\partial T}{\partial V}\right)_S = -\left(\frac{\partial p}{\partial S}\right)_V \tag{1.98}$$

holds.

Indeed, comparing (1.96) and (1.97), we have

$$T = \left(\frac{\partial U}{\partial S}\right)_V, \quad -p = \left(\frac{\partial U}{\partial V}\right)_S.$$

Hence the left-hand side of (1.98) is

$$\left(\frac{\partial}{\partial V}\left(\frac{\partial U}{\partial S}\right)_V\right)_V,$$

which is, by the Schwarz theorem,

$$\left(\frac{\partial}{\partial S}\left(\frac{\partial U}{\partial V}\right)_S\right)_V.$$

This last expression is $\left(\frac{\partial}{\partial S}(-p)\right)_V$ = right-hand side of (1.98).

1.6 Cauchy-Riemann equation

Although we defined analyticity of a function in §1.1 rather naively and formally by (1.1), analyticity in a domain is a very strong condition and it entails many of its intrinsic properties. We describe analyticity in terms of real functions.

Theorem 1.11. *A function $w = f(z) = u + iv$ is analytic in the domain $D \subset \mathbb{C} \iff u,\ v$ are totally differentiable on $D \subset \mathbb{R}^2$ and satisfy the* **Cauchy-Riemann equation**

$$\frac{\partial u}{\partial x} = \frac{\partial v}{\partial y},\ \frac{\partial u}{\partial y} = -\frac{\partial v}{\partial x}; \qquad u_x = v_y,\ u_y = -v_x. \tag{1.99}$$

In this case, we have

$$f'(z) = \frac{\partial f}{\partial x} = u_x + iv_x. \tag{1.100}$$

(Also, substituting the Cauchy-Riemann equation in this formula, we may express it as $f'(z) = \frac{\partial f}{\partial (iy)} = \frac{1}{i}(u_y + iv_y) = v_y - iu_y$.)

Proof. ($\Rightarrow$) Since f is differentiable at each point z_0, substituting the value $\alpha = f'(z_0) = P + iQ$ in (1.1) and comparing the real and imaginary parts, we obtain

$$\begin{aligned} u(x,y) - u(x_0,y_0) &= P(x - x_0) - Q(y - y_0) + o(|z - z_0|) \\ &= (P, -Q)\begin{pmatrix} x - x_0 \\ y - y_0 \end{pmatrix} + o(|z - z_0|) \end{aligned}$$

and

$$\begin{aligned} v(x,y) - v(x_0,y_0) &= Q(x - x_0) + P(y - y_0) + o(|z - z_0|) \\ &= (Q, P)\begin{pmatrix} x - x_0 \\ y - y_0 \end{pmatrix} + o(|z - z_0|). \end{aligned}$$

Hence,

$$\begin{aligned} u(\mathbf{z}) - u(\mathbf{z}_0) &= (P, -Q)(\mathbf{z} - \mathbf{z}_0) + o(|\mathbf{z} - \mathbf{z}_0|), \\ v(\mathbf{z}) - v(\mathbf{z}_0) &= (Q, P)(\mathbf{z} - \mathbf{z}_0) + o(|\mathbf{z} - \mathbf{z}_0|), \end{aligned}$$

where $\mathbf{z} = \begin{pmatrix} x \\ y \end{pmatrix}$, $\mathbf{z}_0 = \begin{pmatrix} x_0 \\ y_0 \end{pmatrix}$ are the real vectors corresponding to $z = x + iy$, $z_0 = x_0 + iy_0$. This means that $u,\ v$ are both totally differentiable at z_0 and $P = u_x,\ Q = -u_y;\quad Q = v_x,\ P = v_y$ holds. Hence, in particular, it follows that $u_x = P = v_y$, $u_y = -Q = -v_x$ and the Cauchy-Riemann equation holds.

($\Leftarrow$) We may trace back the above proof in the reverse direction. With $P = u_x = v_y$, $Q = -u_y = v_x$, $\boldsymbol{h} = \begin{pmatrix} h_1 \\ h_2 \end{pmatrix} \to \mathbf{o}$ ($\Leftrightarrow h := h_1 + ih_2 \to 0$), we have

$$u(x + h_1, y + h_2) - u(x, y) = (u_x, u_y)\,\boldsymbol{h} + o(|\boldsymbol{h}|) = Ph_1 - Qh_2 + o(|h|),$$

$$v(x + h_1, y + h_2) - v(x, y) = (v_x, v_y)\,\boldsymbol{h} + o(|\boldsymbol{h}|) = Qh_1 + Ph_2 + o(|h|),$$

so that

$$f(z + h) - f(z) = Ph_1 - Qh_2 + i(Qh_1 + Ph_2) + o(|h|)$$

$$= (P + iQ)(h_1 + ih_2) + o(|h|).$$

Hence it follows that f is differentiable and $f'(z) = u_x + iv_x$. □

Remark 1.7. (i) To remember the Cauchy-Riemann equation, notice the alphabetical order of letters in the variables $w = u + iv$, $z = x + iy$ and remember first $u_x = v_y$, which is in alphabetical order and then changing the order to replace the real and imaginary parts, then we have a sign change to get $u_y = -v_x$. Also mnemonics for $f'(z)$ is first to think of $f(z)$ as a formal sum $u + iv$ of two real functions and to partially differentiate it with respect to x: to find $\frac{\partial f}{\partial x}$.
(ii) The total differentiability in Theorem 1.11 is assured by the continuity of $\frac{\partial u}{\partial x}, \ldots, \frac{\partial v}{\partial y}$, which is then assured roughly by their differentiability. Thus, if assume that u, v are of C^2-class, then we may appeal to Theorem 1.11.

Corollary 1.5. (Newton-Leibniz rule) *Suppose $f(z)$ is analytic in the domain D and $f'(z) \equiv 0$ (identically 0). Then $f(z)$ must be a constant.*

Proof. By (1.100) and (1.99) we have $u_x = u_y = v_x = v_y = 0$, whence f is a constant. □

Fig. 1.6 Bernhard Riemann, 1826-1866

It follows from Theorem 1.11 that

$$\frac{\partial}{\partial z} f := \left(\frac{\partial}{\partial x} + i\frac{\partial}{\partial y}\right) f = \frac{\partial f}{\partial x} - \frac{\partial f}{\partial (iy)} = u_x - v_y + i(u_y + v_x). \quad (1.101)$$

Symbolically we may write

$$\frac{\partial}{\partial(x+iy)} = \frac{\partial}{\partial z}, \quad \frac{\partial}{\partial(x-iy)} = \frac{\partial}{\partial \bar{z}}. \quad (1.102)$$

Hence

Theorem 1.12. *Suppose $P, Q \in C^2(D)$, where $D \subset \mathbb{R}^2$ is a domain. Then $f(z) = P + iQ$ is analytic in $D \subset \mathbb{C}$ if and only if*

$$\frac{\partial}{\partial z} f = 0. \quad (1.103)$$

This follows from the fact that (1.103) is equivalent to the Cauchy-Riemann equation.

Example 1.20. (2-dimensional irrotational flow) Cf. §1.5.2. In the case of a 2-dimensional irrotational flow, the **stream function** ψ is given as the imaginary part of the analytic function $w = \varphi + i\psi$ (the complex potential) and the **stream line** is given by $d\psi = 0$, i.e. $\psi =$ constant while the **equi-potential line** by ($d\varphi = 0$). The Cauchy-Riemann equation indicates the orthogonality of the equi-potential line and the stream line.

Proposition 1.3. *In the polar coordinates with $r \neq 0$, the Cauchy-Riemann equation $u_x = v_y, \quad u_y = -v_x$ reads*

$$u_r = \frac{1}{r} v_\theta, \quad v_r = -\frac{1}{r} u_\theta. \tag{1.104}$$

The derivative is given by

$$w' = u_r \cos\theta + v_r \sin\theta + i(v_r \cos\theta - u_r \sin\theta). \tag{1.105}$$

Proof. We may express the polar coordinates as a vector-valued function: For $\boldsymbol{x} = \begin{pmatrix} r \\ \theta \end{pmatrix}$, we have

$$\begin{pmatrix} x \\ y \end{pmatrix} = \boldsymbol{y} = \boldsymbol{f}(\boldsymbol{x}) = \begin{pmatrix} r\cos\theta \\ r\sin\theta \end{pmatrix}.$$

By Definition 1.4

$$\begin{pmatrix} \mathrm{d}x \\ \mathrm{d}y \end{pmatrix} = \mathrm{d}\boldsymbol{f} = \nabla \boldsymbol{f} \mathrm{d}\boldsymbol{x} = \nabla \boldsymbol{f} \begin{pmatrix} \mathrm{d}r \\ \mathrm{d}\theta \end{pmatrix},$$

with gradient of $\boldsymbol{f}$ given by

$$\nabla \boldsymbol{f} = \begin{pmatrix} \cos\theta & -r\sin\theta \\ \sin\theta & r\cos\theta \end{pmatrix},$$

so that the Jacobian is

$$\frac{\partial(x, y)}{\partial(r, \theta)} = r.$$

Hence (in view of $r \neq 0$) by Cramér's rule,

$$\begin{pmatrix} \mathrm{d}r \\ \mathrm{d}\theta \end{pmatrix} = (\nabla \boldsymbol{f})^{-1} \mathrm{d}\boldsymbol{x} = \begin{pmatrix} \cos\theta & \sin\theta \\ -\frac{1}{r}\sin\theta & \frac{1}{r}\cos\theta \end{pmatrix} \begin{pmatrix} \mathrm{d}x \\ \mathrm{d}y \end{pmatrix},$$

or

$$\frac{\partial r}{\partial x} = \cos\theta, \quad \frac{\partial r}{\partial y} = \sin\theta, \quad \frac{\partial \theta}{\partial x} = -\frac{1}{r}\sin\theta, \quad \frac{\partial \theta}{\partial y} = \frac{1}{r}\cos\theta. \tag{1.106}$$

Now by the chain rule, the Cauchy-Riemann equation reads

$$\begin{pmatrix} \frac{\partial r}{\partial x} & \frac{\partial \theta}{\partial x} \\ \frac{\partial r}{\partial y} & \frac{\partial \theta}{\partial y} \end{pmatrix} \begin{pmatrix} \frac{\partial u}{\partial r} \\ \frac{\partial u}{\partial \theta} \end{pmatrix} = \begin{pmatrix} \frac{\partial u}{\partial x} \\ \frac{\partial u}{\partial y} \end{pmatrix} = \begin{pmatrix} \frac{\partial v}{\partial y} \\ -\frac{\partial v}{\partial x} \end{pmatrix} = \begin{pmatrix} \frac{\partial \theta}{\partial y} & -\frac{\partial r}{\partial y} \\ -\frac{\partial \theta}{\partial x} & \frac{\partial r}{\partial x} \end{pmatrix} \begin{pmatrix} \frac{\partial v}{\partial \theta} \\ -\frac{\partial v}{\partial r} \end{pmatrix}. \tag{1.107}$$

Hence substituting (1.106) and solving in $\begin{pmatrix} \frac{\partial u}{\partial r} \\ \frac{\partial u}{\partial \theta} \end{pmatrix}$, we obtain

$$\begin{pmatrix} \frac{\partial u}{\partial r} \\ \frac{\partial u}{\partial \theta} \end{pmatrix} = \begin{pmatrix} \cos\theta & \sin\theta \\ -r\sin\theta & r\cos\theta \end{pmatrix} \begin{pmatrix} \frac{1}{r}\cos\theta & -\sin\theta \\ \frac{1}{r}\sin\theta & \cos\theta \end{pmatrix} \begin{pmatrix} \frac{\partial v}{\partial \theta} \\ -\frac{\partial v}{\partial r} \end{pmatrix} \tag{1.108}$$

$$= \begin{pmatrix} \frac{1}{r} & 0 \\ 0 & r \end{pmatrix} \begin{pmatrix} \frac{\partial v}{\partial \theta} \\ -\frac{\partial v}{\partial r} \end{pmatrix},$$

which is (1.104).

For (1.105), we have

$$u_x = u_r \cos\theta - \frac{1}{r} u_\theta \sin\theta, \; v_x = v_r \cos\theta - \frac{1}{r} v_\theta \sin\theta \tag{1.109}$$

whence by (1.104), we deduce (1.105), which completes the proof. □

Example 1.21. (Complex velocity) Cf. §1.5.2. We consider the **complex velocity** corresponding to the velocity vector $\boldsymbol{v}$ in the form of the complex conjugate

$$\bar{v} = P - iQ. \tag{1.110}$$

Then

$$i\frac{\partial}{\partial \bar{z}}\bar{v} = i\left(\frac{\partial}{\partial x} - i\frac{\partial}{\partial y}\right)(P - iQ) = \frac{\partial Q}{\partial x} - \frac{\partial P}{\partial y} + i\left(\frac{\partial P}{\partial x} + \frac{\partial Q}{\partial y}\right). \tag{1.111}$$

Comparing (1.111) with (1.85), we conclude that

$$W = \operatorname{curl} \boldsymbol{v} + i \operatorname{div} \boldsymbol{v} = i\frac{\partial}{\partial \bar{z}}\bar{v}. \tag{1.112}$$

It follows from Theorem 1.4 that if $\frac{\partial}{\partial \bar{z}}\bar{v}$ is analytic in a domain D with a closed curve C as its boundary, then

$$\int_C (\operatorname{curl} \boldsymbol{v} + i \operatorname{div} \boldsymbol{v})\, \mathrm{d}z = i \int_C \frac{\partial}{\partial \bar{z}}\bar{v}\, \mathrm{d}z = 0, \tag{1.113}$$

whence that the fluid is incompressible and irrotational.

However, if it has a pole in D, as in the case of the **vortex filament** $i\frac{k}{z}$, where $k > 0$ is a constant, then the integral along the curve C encircling the pole is not zero but the residue at its pole as furnished by Theorem 1.7. In the case of the vortex filament, the residue at the origin is $-2\pi k$ and (1.113) reads

$$\int_C \operatorname{curl} \boldsymbol{v}\, \mathrm{d}z + i \int_C \operatorname{div} \boldsymbol{v}\, \mathrm{d}z = -2\pi k, \tag{1.114}$$

whence $\int_C \operatorname{curl} \boldsymbol{v}\, \mathrm{d}z = -2\pi k \neq 0$, i.e. there is a whirl at the pole. For a higher point of view on this from the Sato hyperfunction, cf. [Imai (1963)].

Exercise 1.8. Find an analytic function $w = f(z) = u + iv$ in $z = x + iy$ such that u does not depend on θ.

Exercise 1.9. View the following complex functions $u + iv = w = f(z) \Longleftrightarrow u = u(z)$, $v = v(z)$ as a vector-valued function (cf. (2.48)):

$$\begin{pmatrix} u \\ v \end{pmatrix} = \boldsymbol{w} = \boldsymbol{f}(\boldsymbol{z}) = \begin{pmatrix} u(\boldsymbol{z}) \\ v(\boldsymbol{z}) \end{pmatrix} = \begin{pmatrix} u(x,y) \\ v(x,y) \end{pmatrix}$$

in $\boldsymbol{z} = \begin{pmatrix} x \\ y \end{pmatrix}$ and find u_x, u_y, v_x, v_y, the total differential $\mathbf{d}f$ and the Jacobian

$$\frac{\partial(u,v)}{\partial(x,y)} = \begin{vmatrix} \frac{\partial u}{\partial x} & \frac{\partial u}{\partial y} \\ \frac{\partial v}{\partial x} & \frac{\partial v}{\partial y} \end{vmatrix}. \tag{1.115}$$

Also check if the inverse function exists.

(i) $w = z^3$ (ii) $w = \frac{1}{z}$, $z \neq 0$, in all subsequent problems, the denominators do not vanish. (iii) $w = \frac{z}{1+z}$ (iv) $w = \frac{z+2}{z+1}$ (v) $w = z + \frac{1}{z}$ (vi) $w = ze^z$ (vii) $w = \frac{z^2+1}{2z}$.

In the following (ix), (x), we use the polar coordinates

$$r = \sqrt{x^2 + y^2}, \; x = r\cos\theta, \; y = r\sin\theta.$$

(ix) $w = \log r + i\theta$ (x) $w = \sqrt{r}\cos\frac{\theta}{2} + i\sqrt{r}\sin\frac{\theta}{2}$.

Example 1.22. (i) The complex function

$$u + iv = w = z^2 = (x + iy)^2 = x^2 - y^2 + i(2xy), \tag{1.116}$$

is equivalent to the vector-valued function

$$\begin{cases} u = x^2 - y^2 \\ v = 2xy, \end{cases} \tag{1.117}$$

whence

$$u_x = 2x, \; u_y = -2y, \; v_x = 2y, \; u_y = 2x. \tag{1.118}$$

Hence the Cauchy-Riemann equation holds. Since u, v are totally differentiable, w is analytic in the whole domain $\mathbb{C}$ and $f'(z) = u_x + iv_x = 2x + i2y = 2z$, i.e. $(z^2)' = 2z$.

(ii) The complex function

$$u + iv = w = \log\sqrt{x^2 + y^2} + i\arctan\frac{y}{x}, \tag{1.119}$$

for $x > 0$ is equivalent to the vector-valued function

$$\begin{cases} u = \log\sqrt{x^2+y^2} \\ v = \arctan\dfrac{y}{x}, \end{cases} \tag{1.120}$$

whence

$$u_x = \frac{x}{x^2+y^2},\ u_y = \frac{y}{x^2+y^2}\ v_x = -\frac{y}{x^2+y^2},\ u_y = \frac{x}{x^2+y^2}. \tag{1.121}$$

Hence the Cauchy-Riemann equation holds. Since u, v are totally differentiable, w is analytic in the domain $\operatorname{Re} z > 0$ and $f'(z) = u_x + iv_x = \frac{x}{x^2+y^2} - i\frac{y}{x^2+y^2} = \frac{x-iy}{x^2+y^2} = \frac{1}{z}$ in $\operatorname{Re} z > 0$.

I.e. we have deduced the important formula

$$(\log z)' = \frac{1}{z}, \quad \operatorname{Re} z > 0. \tag{1.122}$$

The Jacobian is $\frac{\partial(u,v)}{\partial(x,y)} = \frac{x^2+y^2}{(x^2+y^2)^2} = \frac{y}{x^2+y^2} = \frac{1}{|z|^2}$. This is in conformity with (1.122).

For more details, cf. Example 1.24.

Exercise 1.10. Check whether the following functions are analytic functions in $z = x + iy$.
(i) $w = 2x + ixy^2$ (ii) $w = \frac{x-iy}{x^2+y^2}$ (iii) $w = e^x e^{-iy}$ (iv) $w = \sin x \cosh y + i \cos x \sinh y$.

Solution. (iv) By (1.24), we deduce that $w = \sin x \cos iy + i \cos x \sin iy = \sin(x + iy) = \sin z$.

Exercise 1.11. Determine the real constants $a, b, c, d, e \in \mathbb{R}$ so that the following functions w become analytic functions in $z = x + iy$.
(i) $w = x + ay + i\,(bx + cy)$
(ii) $w = x^2 + axy + by^2 + i\,(cx^2 + dxy + ey^2)$
(iii) $w = \cos x \cosh y + a \cos x \sinh y + i\,(b \sin x \sinh y + \sin x \cosh y)$.

Solution. (i) $w = x + ay + i(-ax + y)$.
(ii)

$$w = x^2 + axy - y^2 + i\left(-\frac{1}{2}ax^2 + 2xy + \frac{1}{2}ay^2\right), \tag{1.123}$$

where $a \in \mathbb{R}$.
(iii)

$$u = \cos x(\cosh y + a \sinh y), \quad v = \sin x(b \sinh y + \cosh y). \tag{1.124}$$

By the Cauchy-Riemann equation, we must have

$$w = \cos x(\cosh y - \sinh y) + i \sin x(-\sinh y + \cosh y). \tag{1.125}$$

Simplifying thereby using (1.24)

$$w = (\cosh y - \sinh y)(\cos x + i \sin x) = e^{ix}e^{-y} = e^{x+iy} = e^{iz}. \tag{1.126}$$

Example 1.23. By Exercise 1.11 (i), $w = x$, $w = x - iy = \bar{z}$, $w = 2iy = z - \bar{z}$ are not analytic functions. Neither is $w = \frac{x+iy}{x^2+y^2} = \frac{z}{z\bar{z}} = \frac{1}{\bar{z}}$.
By Exercise 1.11 (ii) $w = x^2 + iy^2$, $w = x^2 + y^2 = |z|^2$, $w = x^2 + y^2 + 2ixy$ are not analytic functions.

1.7 Inverse functions

We recall the inverse function theorem for functions several (real) variables.

Theorem 1.13. (Inverse function theorem) *If a vector-valued function* $\boldsymbol{w}$ *is differentiable at a point* $\boldsymbol{z}_0$ *and* $J_{\boldsymbol{w}}(\boldsymbol{z}_0) \neq 0$ *in the neighborhood, then there exists a local inverse function of* $\boldsymbol{w}$ *and is locally* C^1.

Corollary 1.6. (Inverse function theorem for complex variables) *If a function* $w = f(z)$ *is analytic in a domain* D *and at* $z_0 \in D$, $f'(z_0) \neq 0$, *then the inverse function* $g(w) = f^{-1}(w)$ *exists in the neighborhood of* $w_0 = f(z_0)$ *and is analytic at* w_0, *with the derivative given by*

$$g'(w) = \frac{1}{f'(z)}, \quad \frac{\mathrm{d}z}{\mathrm{d}w} = \left(\frac{\mathrm{d}w}{\mathrm{d}z}\right)^{-1}.$$

Proof. When we view the complex function function $u + iv = w = f(z)$ as the vector-valued function $\boldsymbol{w} = \begin{pmatrix} u \\ v \end{pmatrix}$, $u = u(\boldsymbol{z})$, $v = v(\boldsymbol{z})$, $\boldsymbol{z} = \begin{pmatrix} x \\ y \end{pmatrix}$, then the Jacobian $J_{\boldsymbol{w}} = \frac{\partial(u,v)}{\partial(x,y)}$ is $|f'(z)|^2$:

$$J(x, y) = J_{\boldsymbol{w}}(\boldsymbol{z}) = |f'(z)|^2. \tag{1.127}$$

Hence we may appeal to the inverse function theorem. □

Example 1.24. The (complex) **logarithm function** $\log w$ (which is often denoted $\ln z$) is the inverse function of the exponential function $w = e^z$. In view of the periodicity (1.17), it is a multi-valued function.

$$\log w = z = x + iy = \log |w| + i \arg w = \log |w| + i \arctan \frac{v}{u}, \tag{1.128}$$

where the last expression is valid for $u \neq 0$. In practice, we want to treat a multi-valued function as an ordinary function. For this we restrict the range of the argument to length 2π and we call each single-valued function a branch. For any branch, we have $(\log w)' = \frac{1}{w}$, $\quad w \neq 0$. We note that the multi-valuedness of the arctan function corresponds to that of the logarithm function.

Indeed, the Jacobian is $J(x,y) = |e^z|^2 = e^{2x} \neq 0$, so that the inverse function exists. To prove (1.128), we write

$$u + iv = w = e^z = e^x e^{iy}.$$

Then we have

$$\begin{cases} u = e^x \cos y \\ v = e^x \sin y \end{cases} \iff z = \log w$$

or $|w| = e^x$ and $\tan y = \frac{v}{u}$ for $u \neq 0$, whence

$$x = \log |w| \quad \text{and} \quad y = \arctan \frac{v}{u}.$$

Since $\arctan \frac{v}{u}$ is nothing but $\arg w$, we conclude (1.128).

We often choose the **principal branch** $\log z$ (sometimes written with capital L) of the logarithm, where the principal branch means that the argument lies between $-\pi$ and π:

$$\log z = \log |z| + i \arg z, \quad -\pi < \arg z \leq \pi. \tag{1.129}$$

Of course, we can consider any branch, e.g. the one with $0 \leq \arg z < 2\pi$ will be used in Theorem 1.22.

For the complex variable $s = \sigma + it$ and $n \in \mathbb{N}$, we define the **power function** a^z for a non-zero complex number a by

$$a^z = e^{z \log a}, \tag{1.130}$$

where a suitable branch is to be taken for $\log a$. In particular, for a natural number n and $s = \sigma + it$ we define n^{-s} by

$$n^{-s} = e^{-s \log n} = n^{-\sigma}(\cos(t \log n) - i \sin(t \log n)), \tag{1.131}$$

with $\log n$ designating the real logarithm. With this one can introduce the celebrated Riemann zeta-function as the Dirichlet series

$$\sum_{n=1}^{\infty} \frac{1}{n^s} \tag{1.132}$$

which converges absolutely for $\sigma > 1$ and defines an analytic function.

Exercise 1.12. Deduce the following expressions for the inverse trigonometric functions

$$\arcsin z = \frac{1}{i}\log\left(iz + \sqrt{1-z^2}\right), \quad \arctan z = \frac{1}{2i}\log\frac{i-z}{i+z} \quad (z \neq \pm i). \tag{1.133}$$

Example 1.25. (2-dimensional flow of a vortex) Consider the complex potential given by $w = -\frac{\Gamma}{2\pi i}\log z$, where $\Gamma > 0$ is a constant. Putting $w = \varphi + i\psi$ and $z = re^{i\vartheta}$, we have by (1.128)

$$\varphi = -\frac{\Gamma}{2\pi}\vartheta, \quad \psi = \frac{\Gamma}{2\pi}\log r. \tag{1.134}$$

The stream line $\psi = \text{const.}$ is $r = \text{const.}$, i.e. the concentric circles around the origin. This is the flow (in the positive direction) around the vortex at the origin as we see presently. The radiation velocity $v(r)$, the tangential velocity $v(\vartheta)$ are given respectively by

$$v(r) = -\frac{\partial\varphi}{\partial r}, \quad v(\vartheta) = -\frac{1}{r}\frac{\partial\varphi}{\partial\vartheta}.$$

Hence, in our case, by (1.134)

$$v(r) = 0, \quad v(\vartheta) = \frac{\Gamma}{2\pi r}, \tag{1.135}$$

so that the **induced velocity** caused by the vortex is only the circulation $v(\vartheta)$ around the origin. The constant Γ turns out to be the circulation and is called the **intensity** of the flow.

Apparently, if the vortex is at the point α, the complex potential is given by

$$w = -\frac{\Gamma}{2\pi i}\log(z-\alpha). \tag{1.136}$$

Example 1.26. (The Karman vortex row) First consider the case of a **vortex row** in which there are infinitely many vortices with the same intensity and rotating in the positive direction at distance a apart. We consider the $2n+1$ vortices with the central one at the origin. Then by (1.136) and the principle of superposition, the complex potential is

$$w_n = -\frac{\Gamma}{2\pi i}\left(\log z + \sum_{k=1}^{n}(\log(z-ka) + \log(z+ka))\right). \tag{1.137}$$

For the subsequent argument, cf. §1.11.3. Recall the partial fraction expansion of the cotangent function (1.231) valid for all $z \in \mathbb{C}$ save for multiples of π:

$$\begin{aligned}\cot z &= \frac{1}{z} + \sum_{n=-\infty}^{\infty}{}' \left(\frac{1}{z-\pi n} + \frac{1}{\pi n}\right) \\ &= \frac{1}{z} + 2z \sum_{n=1}^{\infty} \frac{1}{z^2 - \pi^2 n^2},\end{aligned} \tag{1.138}$$

where the prime on the summation sign means that the term with $n = 0$ is omitted. Since the series is uniformly convergent in any domain not containing multiples of π, we may integrate the first equality of (1.138) term by term from 1 to z to obtain

$$\log \sin z = \log z + \sum{}' \log(z - \pi n) e^{\frac{z}{\pi n}} = \log z + \sum_{n=1}^{\infty} \log(z^2 - \pi^2 n^2), \tag{1.139}$$

on summing the infinite series in pairs n and $-n$.

Hence the limit $w = \lim_{n\to\infty} w_n$ of (1.137) may be interpreted to mean

$$\begin{aligned} w &= -\frac{\Gamma}{2\pi i}\left(\log z + \sum{}' \log(z - an) e^{\frac{z}{an}}\right) \\ &= -\frac{\Gamma}{2\pi i}\left(\log z + \sum_{n=1}^{\infty} \log(z^2 - a^2 n^2)\right)\end{aligned} \tag{1.140}$$

which amounts to

$$w = -\frac{\Gamma}{2\pi i} \log \sin \frac{\pi z}{a}. \tag{1.141}$$

The above argument gives a legitimation in Exercise 1.13 below. Note that we may interpret (1.139) to mean

$$\sin z = z \prod_{n=-\infty}^{\infty}{}' \left(1 - \frac{z}{\pi n}\right) e^{\frac{z}{\pi n}}, \tag{1.142}$$

which leads to Theorem 1.21 on summing in pairs.

With (1.141) in mind, we may consider the **Karman vortex street** which is the row of two rows of infinite vortices with same intensity but with the opposite rotation. We express them as two rows above and below the real axis at the height h and each vortex in the one row lies in the middle of the other row. The complex potential of the above vortex row is given by

$$w = -\frac{\Gamma}{2\pi i} \log \sin \frac{\pi}{a}\left(z - \frac{ih}{2}\right),$$

and that of the lower row is

$$w = -\frac{\Gamma}{2\pi i} \log \sin \frac{\pi}{a}\left(z + \frac{ih}{2} + \frac{a}{2}\right).$$

Hence their composition potential is

$$w = -\frac{\Gamma}{2\pi i} \log \frac{\sin \frac{\pi}{a}\left(z - \frac{ih}{2}\right)}{\cos \frac{\pi}{a}\left(z + \frac{ih}{2}\right)}. \tag{1.143}$$

The stream function ψ is the imaginary part of (1.143) and can be computed to be

$$\psi = \frac{\Gamma}{4\pi} \log \frac{\cosh \frac{2\pi}{a}\left(y - \frac{h}{2}\right) - \cos \frac{2\pi}{a} x}{\cosh \frac{2\pi}{a}\left(y + \frac{h}{2}\right) + \cos \frac{2\pi}{a} x}. \tag{1.144}$$

It was known due to Karman that the vortex street is stable when $\sin \frac{\pi h}{a} = 1$ or $\frac{h}{a} \approx 0.281$.

Let ν be the **kinematic coefficient of viscosity** of the fluid, which is the ratio of the **coefficient of viscosity** to the density of the fluid. Let r be the radius of the cylinder and let U be the uniform speed of the flow. Then the **Reynolds number** R_e is defined by

$$R_e = \frac{2rU}{\nu}. \tag{1.145}$$

It was shown by Roshko (1953) that the Karman vortex street arises for $R_e \geq 40$.

Exercise 1.13. In most of the textbooks on fluid mechanics, one finds the following argument: Since $\log(z - ka) + \log(z + ka) = \log(z^2 - (ka)^2)$, it follows, after slight transformation, that

$$w_n = -\frac{\Gamma}{2\pi i}\left(\log \frac{\pi z}{a} \prod_{k=1}^{n}\left(1 - \frac{z^2}{k^2 a^2}\right)\right) + \text{const.}$$

Hence choosing the coordinates which annihilate the constant and letting $n \to \infty$, we obtain

$$w = \lim_{n\to\infty} w_n = -\frac{\Gamma}{2\pi i}\left(\log \frac{\pi z}{a} \prod_{k=1}^{\infty}\left(1 - \frac{z^2}{k^2 a^2}\right)\right). \tag{1.146}$$

Comparing (1.146) with (1.142), we conclude (1.141).
Find fallacies of the argument.
Deduce (1.141) using the second equality of (1.138).

Exercise 1.14. Deduce (1.144).

Solution. In view of the form of (1.143), the imaginary part can be computed from $\log|\frac{\sin z}{\cos z}|$ and it suffices to find $|\sin(x+iy)|$. Since

$$\sin(x+iy) = \sin x \frac{e^{-y}+e^{y}}{2} + \cos x \frac{e^{-y}-e^{y}}{2i},$$

it follows that

$$\begin{aligned} |\sin(x+iy)|^2 &= (\sin^2 x + \cos^2 x)\left(\frac{e^{2y}+e^{-2y}}{4}\right) + \frac{1}{2}(\sin^2 x - \cos^2 x) \\ &= \frac{1}{2}(\cosh 2y - \cos 2x). \end{aligned}$$

Similarly,

$$|\cos(x+iy)|^2 = \frac{1}{2}(\cosh 2y + \cos 2x).$$

Equation (1.144) immediately follows from these.

1.8 Around Jensen's formula

In Theorem 1.9, we specialize C to be a circle $z = z_0 + re^{i\theta}$, $0 \le \theta \le 2\pi$. Then the following theorem is immediate.

Theorem 1.14. (Gauss' mean value theorem) *If $f(z)$ is analytic inside the circle $C: z = z_0 + re^{i\theta}, \quad \theta \in [0, 2\pi]$, then*

$$f(z_0) = \frac{1}{2\pi}\int_0^{2\pi} f(z_0 + re^{i\theta})\, d\theta. \tag{1.147}$$

(1.147) holds for harmonic functions, too, for which we refer to Definition 2.8. $u = u(\boldsymbol{z}) = u(x,y)$ is a **harmonic function** if it is a C^2-function satisfying the **Laplace equation**

$$\Delta u = \frac{\partial^2 u}{\partial x^2} + \frac{\partial^2 u}{\partial y^2} = 0. \tag{1.148}$$

Indeed, by Theorem 2.21, the real and imaginary parts of an analytic function are harmonic in the region of analyticity (where we use the convention that $\mathbb{C} \sim \mathbb{R}^2$). Hence taking real and imaginary parts, (1.148) follows.

In particular, suppose $f(z)$ is analytic in a domain D containing a disc $|z - z_0| \le r$ and is free from zeros there. Then (1.147) reads

$$\log|f(z_0)| = \frac{1}{2\pi}\int_0^{2\pi} \log\left|f\left(z_0 + re^{i\theta}\right)\right|\, d\theta. \tag{1.149}$$

Further, we consider the special case of $z_0 = 0$:

$$\log |f(0)| = \frac{1}{2\pi} \int_0^{2\pi} \log \left| f\left(re^{i\theta}\right)\right| \, d\theta. \tag{1.150}$$

We follow [Ahlfors (1979), p. 207] to consider the case where $f(z)$ has a zero $re^{i\theta_0}$ on the circle $C : |z| = r$. For simplicity, we restrict to simple zeros hereafter. We can easily incorporate the case of multiple zeros.

Theorem 1.15. (1.150) *remains true if $f(z)$ has a (simple) zero $re^{i\theta_0}$ on C; or more properly,* (1.150) *is true for $g(z) = \frac{f(z)}{z-re^{i\theta_0}}$.*

Proof. (1.150) for $g(z)$ reads

$$\log |f(0)| - \log \left|re^{i\theta_0}\right| = \frac{1}{2\pi} \int_0^{2\pi} \log \left| \frac{f(re^{i\theta})}{e^{i\theta} - e^{i\theta_0}} \right| \, d\theta - \frac{1}{2\pi} \int_0^{2\pi} \log r \, d\theta,$$

or

$$\log |f(0)| = \frac{1}{2\pi} \int_0^{2\pi} \log \left| \frac{f(re^{i\theta})}{e^{i\theta} - e^{i\theta_0}} \right| \, d\theta.$$

Hence if we show that

$$\int_0^{2\pi} \log \left|e^{i\theta} - e^{i\theta_0}\right| \, d\theta = 0,$$

then we may assert that the above formulas is a generalization of (1.150). Or we are to prove that

$$\int_0^{2\pi} \log \left|1 - e^{i\theta}\right| \, d\theta = 0,$$

which amounts to proving

$$\int_0^{2\pi} \log \sin \frac{\theta}{2} \, d\theta = 0.$$

This is a consequence of one of well-known Euler's integrals

$$\int_0^{\frac{\pi}{2}} \log \sin x dx = -\frac{\pi}{2} \log 2 \tag{1.151}$$

to be discussed below. □

Generalizing the idea of eliminating the zeros, we may prove

Theorem 1.16. (Jensen's formula). *Let $f(z)$ be analytic in the disc $|z| \le r$ and suppose it has zeros at $a_1, a_2, \ldots, a_n$. in $|z| < r$ and that 0 is not a zero of f. Then*

$$\log |f(0)| = -\sum_{j=1}^{n} \log \frac{r}{|a_j|} + \frac{1}{2\pi} \int_0^{2\pi} \log \left|f(re^{i\theta})\right| \, d\theta. \tag{1.152}$$

Proof follows on applying (1.150) to the Blaschke product (cf. (1.289))

$$g(z) = f(z) \prod_{j=1}^{n} \frac{r^2 - \overline{a}_j z}{r(z - a_j)}. \tag{1.153}$$

1.9 Residue calculus again

First we find the value of the definite integral. The method using anti-derivatives will be discussed in §2.6. The reasoning is quite analogous to that of Example 1.36, showing the power of partial fraction expansions.

Example 1.27. Suppose $b, c \in \mathbb{R}$ satisfy $b^2 - 4c < 0$. We find the value of the integral

$$\int_{-\infty}^{\infty} \frac{1}{(x^2 + bx + c)^2} \, dx = \frac{4\pi\sqrt{4c - b^2}}{(4c - b^2)^2}. \tag{1.154}$$

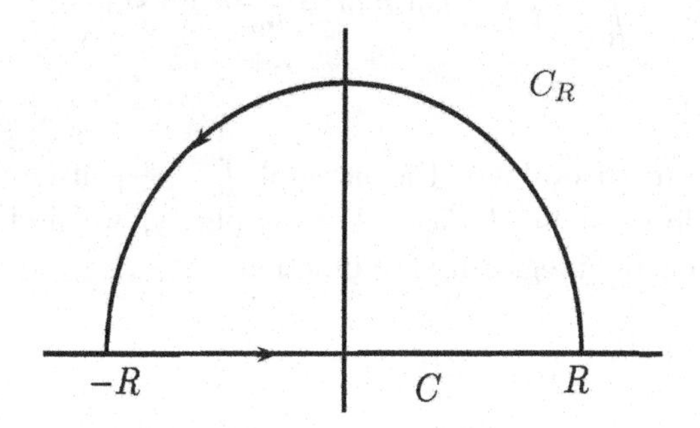

Fig. 1.7 The curve C for Example 1.27

Proof. We integrate the function $f(z) = \frac{1}{(z^2+bz+c)^2}$ along the contour C consisting of the upper semi-circle $C_R : z = Re^{i\theta}$ of radius R going to infinity and the line segment $[-R, R]$. Solving the equation $z^2 + bz + c = 0$, we get the solutions $\alpha, \bar{\alpha}$, where

$$\alpha = \frac{-b + \sqrt{4c - b^2} i}{2}, \tag{1.155}$$

with bar denoting the complex conjugate. The following partial fraction expansion (1.229) will be found in Example 1.36.

$$\frac{1}{(z^2 + bz + c)^2} = \frac{-\frac{1}{4c-b^2}}{(z - \alpha)^2} + \frac{-\frac{2\sqrt{4c-b^2}}{(4c-b^2)^2} i}{z - \alpha} + h(z), \tag{1.156}$$

where $h(z)$ is the holomorphic part at $z = \alpha$ and is given by $\frac{-\frac{1}{4c-b^2}}{(z-\bar{\alpha})^2} + \frac{\frac{2\sqrt{4c-b^2}}{(4c-b^2)^2}i}{z-\bar{\alpha}}$. Since $f(z)$ has only one second-order pole at $z = \alpha$ inside C, we only need the coefficient B of $\frac{1}{z-\alpha}$. By the residue theorem,

$$\int_C f(z)\,\mathrm{d}z = 2\pi i B = \frac{4\pi\sqrt{4c-b^2}}{(4c-b^2)^2}.$$

On the other hand,

$$\int_C f(z)\,\mathrm{d}z = \int_{C_R} f(z)\,\mathrm{d}z + \int_{-R}^{R} f(x)\,\mathrm{d}x \to \int_{-\infty}^{\infty} f(x)\,\mathrm{d}x \tag{1.157}$$

as $R \to \infty$ because $\int_{C_R} f(z)\,\mathrm{d}z = O(R^{-3})$.

Example 1.28. Let $a > 0, b > 0$. We contend that

$$\mathcal{S}[f](a) = \int_0^\infty \frac{t}{t^4+b^4} \sin at\,\mathrm{d}t = \frac{\pi}{2b^2} e^{-\frac{ab}{\sqrt{2}}} \sin\frac{ab}{\sqrt{2}} \tag{1.158}$$

with $f(t) = \frac{t}{t^4+b^4}$.

Proof. We are to evaluate the integral $\int_0^\infty \frac{x}{x^4+1} \sin abx\,\mathrm{d}x$, which is $2b^2$ times the integral in (1.158). For simplicity, we find the value of $\int_0^\infty \frac{x}{x^4+1} \sin ax\,\mathrm{d}x$ by integrating the function

$$f(z) = \frac{z}{z^4+1} e^{iaz}$$

along the contour C in Example 1.27, consisting of the upper semi-circle $C_R : z = Re^{i\theta}$ of radius R going to infinity and the line segment $[-R, R]$. Since $f(z)$ has no pole on the real axis and its order of magnitude at infinity is $O(|z|^{-4})$, analysis is much easier than that in Example 1.30 or in Example 1.13. There are two simple poles α, β inside of C, where

$$\alpha = e^{\frac{2\pi i}{8}} = e^{\frac{\pi i}{4}} = \frac{1+i}{\sqrt{2}},\ \beta = e^{\frac{3\pi i}{4}} = \frac{-1+i}{\sqrt{2}}.$$

The residues at these poles may be found easily as follows on using results from Exercise 2.25, (x'):

$$\operatorname*{Res}_{z=\alpha} f(z) = \lim_{z\to\alpha} \frac{z}{(z-\bar{\alpha})(z^2+\sqrt{2}z+1)} e^{iaz} \tag{1.159}$$

$$= \frac{\alpha}{(\alpha-\bar{\alpha})(\alpha^2-\sqrt{2}\alpha+1+2\sqrt{2}\alpha)} e^{ia\alpha} = -\frac{i}{4} e^{-\frac{a}{\sqrt{2}}} e^{\frac{a}{\sqrt{2}}i}.$$

Similarly,

$$\operatorname*{Res}_{z=\beta} f(z) = \lim_{z\to\beta} \frac{z}{(z-\bar{\beta})(z^2-\sqrt{2}z+1)} e^{iaz} \tag{1.160}$$

$$= \frac{\beta}{(\beta-\bar{\beta})(\beta^2+\sqrt{2}\beta+1-2\sqrt{2}\beta)} e^{ia\beta} = \frac{i}{4} e^{-\frac{a}{\sqrt{2}}} e^{-\frac{a}{\sqrt{2}}i}.$$

It therefore follows that

$$\sum_{z=\alpha,\beta} \operatorname{Res} f(z) = \frac{1}{2} e^{-\frac{a}{\sqrt{2}}} \sin \frac{a}{\sqrt{2}}. \tag{1.161}$$

On the other hand,

$$\int_C f(z)\,\mathrm{d}z = \int_{C_R} f(z)\,\mathrm{d}z + \int_{-R}^{R} f(x)\,\mathrm{d}x \to \int_{-\infty}^{\infty} f(x)\,\mathrm{d}x \tag{1.162}$$

as $R \to \infty$ because $\int_{C_R} f(z)\,\mathrm{d}z = O(R^{-3})$. We note that there is no need to appeal to Jordan's lemma.

Hence comparing (1.161) and (1.162), we conclude that

$$\int_{-\infty}^{\infty} \frac{x}{x^4+1} e^{iax}\,\mathrm{d}x = \pi i e^{-\frac{a}{\sqrt{2}}} \sin \frac{a}{\sqrt{2}} \tag{1.163}$$

or

$$\int_{-\infty}^{\infty} \frac{x}{x^4+1} \sin abx\,\mathrm{d}x = \pi e^{-\frac{ab}{\sqrt{2}}} \sin \frac{ab}{\sqrt{2}}, \tag{1.164}$$

which is $2b^2$ times of the integral in (1.158), completing the proof.

As will be discussed in § 2.13, the left-hand side integral of (1.158) is the Fourier sine transform:

$$\mathcal{S}[f](a) = \frac{\pi}{2b^2} e^{-\frac{ab}{\sqrt{2}}} \sin \frac{ab}{\sqrt{2}}$$

with $f(t) = \frac{t}{t^4+b^4}$.

We give more examples of the Fourier transforms. For more details, cf. §2.13. First we consider the (inverse) Fourier transform $\int_{-\infty}^{\infty} f(t)e^{iat}\,\mathrm{d}t$ of the function $f(t) = \frac{1}{(t^2+bt+c)^2}$ in Example 1.27.

Example 1.29. Let $f(t) = \frac{1}{(t^2+bt+c)^2}$ with $D := b^2 - 4c < 0$. Then its (inverse) Fourier transform is

$$\mathcal{F}[f](a) = \int_{-\infty}^{\infty} \frac{1}{(t^2+bt+c)^2} e^{iat}\,\mathrm{d}t = 2\pi e^{ia\alpha} \left(\frac{2\sqrt{4c-b^2}}{(4c-b^2)^2} + \frac{a}{4c-b^2} \right). \tag{1.165}$$

Putting $a = 0$, we recover the evaluation in Example 1.27. The symbol $\hat{f}$ is also very frequently used for the Fourier transform as well as $\mathcal{F}[f]$.

Proof. The procedure is almost verbatim to the proof of Example 1.27. We integrate the function $f(z) = \frac{1}{(z^2+bz+c)^2}e^{iaz}$ along the same contour C.

Instead of (1.229) in Example 1.36, we use the Laurent expansion (1.156) around $z = \alpha$

$$\frac{1}{(z^2+bz+c)^2} = \frac{-\frac{1}{4c-b^2}}{(z-\alpha)^2} + \frac{-\frac{2\sqrt{4c-b^2}}{(4c-b^2)^2}i}{z-\alpha} + O(1),$$

where the symbol $O(1)$ is synonymous to $h(z)$, the holomorphic part.

We also need the Taylor expansion

$$e^{iaz} = e^{ia\alpha}e^{ia(z-\alpha)} = e^{ia\alpha}\left(1 + ia(z-\alpha) + O((z-\alpha)^2)\right).$$

Multiplying these, we obtain

$$\frac{e^{iaz}}{(z^2+bz+c)^2} = e^{ia\alpha}\left(\frac{-\frac{1}{4c-b^2}}{(z-\alpha)^2} + \frac{-\frac{a}{4c-b^2}i - \frac{2\sqrt{4c-b^2}}{(4c-b^2)^2}i}{z-\alpha}\right) + O(1)$$

and

$$\operatorname*{Res}_{z=\alpha} f(z) = e^{ia\alpha}\left(-\frac{a}{4c-b^2}i - \frac{2\sqrt{4c-b^2}}{(4c-b^2)^2}i\right).$$

Hence

$$\int_C f(z)\,\mathrm{d}z = 2\pi i \operatorname*{Res}_{z=\alpha} f(z) = 2\pi e^{ia\alpha}\left(\frac{2\sqrt{4c-b^2}}{(4c-b^2)^2} + \frac{a}{4c-b^2}\right),$$

and (1.165) follows.

Remark 1.8. The factor in the parentheses in (1.165) may be expressed as $\rho e^{i\theta}$ with $\rho = \frac{\sqrt{|D|+4}}{|D|^{3/2}}$ and θ is the angle. Hence the right-hand side of (1.165) may be expressed as $2\pi\rho e^{i(\alpha a+\theta)}$ and we may view (1.165) as the inverse Fourier transform of the signal $2\pi\rho e^{i(\alpha a+\theta)}$, where α is given in (1.155). This, along with (1.230), indicates that the reason why one uses rational functions in control theory comes from the fact that as long as one considers the signals in the form of an exponential function, its Fourier (or Laplace) transform is a rational function. This suggests a possibility of considering signals given as more elaborate special functions, say Bessel functions, whose degenerate forms are exponential functions. Then their Fourier transforms are corresponding special functions which sometimes degenerate to rational functions. This might explain an optimal choice of the coefficients of the rational function in control theory as an intrinsic structure of the special function in question.

Exercise 1.15. Confirm the following evaluations (for $b > 0$).
(i) $\int_0^\infty \frac{1}{x^4+b^4} \cos ax \, dx = \frac{\pi}{2b^3} \exp\left(-\frac{ab}{\sqrt{2}}\right) \sin\left(\frac{ab}{\sqrt{2}} + \frac{\pi}{4}\right)$.
(ii) $\int_0^\infty \frac{x^2}{x^4+b^4} \cos ax \, dx = \frac{\pi}{2b} \exp\left(-\frac{ab}{\sqrt{2}}\right) \cos\left(\frac{ab}{\sqrt{2}} + \frac{\pi}{4}\right)$.
(iii) $\int_0^\infty \frac{x^3}{x^4+b^4} \sin ax \, dx = \frac{\pi}{2} \exp\left(-\frac{ab}{\sqrt{2}}\right) \cos\left(\frac{ab}{\sqrt{2}}\right)$.

If the integrand has singularities on the real axis, the analysis is more delicate and we make use of Lemma 1.2, which is a full version of Lemma 1.1 above and claim that one can choose upper or lower semi-circles to avoid the singularities.

Lemma 1.2. *Let $f(z)$ be a meromorphic function with a simple pole at $z = a$ and let c_a be a positively-oriented arc*

$$z - a = re^{i\theta}, \quad \theta_1 \le \theta \le \theta_2. \tag{1.166}$$

Then

$$\int_{c_a} f(z) \, dz \to i(\theta_2 - \theta_1) \operatorname*{Res}_{z=a} f(z). \tag{1.167}$$

In particular, let γ_a be the lower semi-circle with center at a of radius r going to 0. Then

$$\int_{\gamma_a} f(z) \, dz \to 2\pi i \frac{1}{2} \operatorname*{Res}_{z=a} f(z). \tag{1.168}$$

If γ_a' is the upper semi-circle with center at a of radius r going to 0:

$$z - a = re^{-i\theta}, \quad -\pi \le \theta \le 0. \tag{1.169}$$

Then

$$\int_{\gamma_a'} f(z) \, dz \to -2\pi i \frac{1}{2} \operatorname*{Res}_{z=a} f(z). \tag{1.170}$$

Proof. Since

$$\int_{c_a} f(z) \, dz = i \int_{\theta_1}^{\theta_2} f(a + re^{i\theta}) re^{i\theta} \, d\theta \tag{1.171}$$

and by assumption

$$f(a + re^{i\theta}) = \frac{\operatorname{Res}_{z=a} f(z)}{re^{i\theta}} + g(z), \tag{1.172}$$

where $g(z) \to$ *const.* as $r \to 0$, Eqn. (1.167) follows. (1.168) is the case where $\theta_1 = \pi$, $\theta_2 = 2\pi$ while (1.170) is the case where $\theta_1 = -\pi$, $\theta_2 = 0$. □

In Examples 1.28, and 1.29, the vanishing of the integral along the circle of radius $R \to \infty$ as well as the convergence of the infinite integral are clear because of the high order of magnitude of the denominator. When the vanishing is not immediately clear and there is the exponential factor e^{iaz}, $a > 0$, we often have recourse to the following Jordan's lemma. As long as (1.173) holds true on the large circle, then we can just go along in the same way as above and calculation amounts to finding residues. Here we stress the **positivity of the parameter** a.

Lemma 1.3. (Jordan's lemma) *Suppose $f(z)$ is analytic on the circle C : $z = Re^{i\theta}$, where θ runs through that part of the interval $[0, 2\pi]$ as alluded to below and that*

$$f(Re^{i\theta}) = o(1), \quad R \to \infty. \tag{1.173}$$

Suppose $a > 0$.
(i) *If C_1 is a part of the upper semi-circle, then*

$$\int_{C_1} e^{iaz} f(z)\,\mathrm{d}z = o(1). \tag{1.174}$$

If C_1 is a part of the lower semi-circle, then

$$\int_{C_1} e^{-iaz} f(z)\,\mathrm{d}z = o(1). \tag{1.175}$$

(ii) *If C_2 is a part of the left semi-circle plus that part of C with* $\mathrm{Re}\, z < c$, *where $0 < c < R$, then*

$$\int_{C_2} e^{az} f(z)\,\mathrm{d}z = o(1). \tag{1.176}$$

(iii) *Let C_2' be the reflection image of C_2 with respect to the imaginary axis. Then the analogue of* (1.174) *holds:*

$$\int_{C_2'} e^{-az} f(z)\,\mathrm{d}z = o(1). \tag{1.177}$$

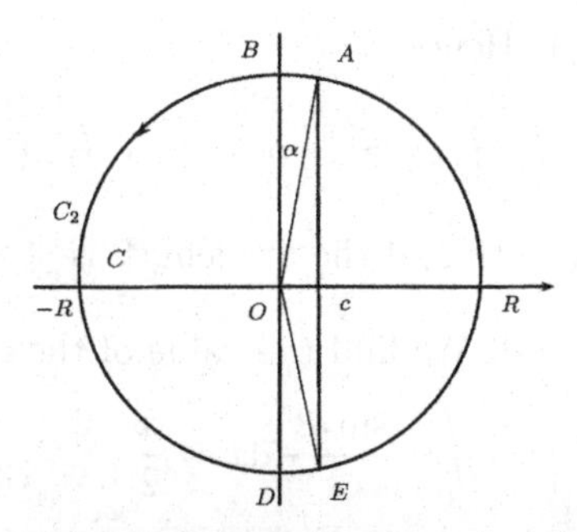

Fig. 1.8 The domain for Lemma 1.3, (ii)

Proof. The case is fundamental when the curve is the first quadrant C : $z = Re^{i\theta}$, $0 \le \theta \le \frac{\pi}{2}$ of the circle.
(i) By (1.173) and Jordan's inequality (2.99), the integral on the left-side of (1.174) is

$$\left|\int_{C_1} e^{iaz} f(z)\,\mathrm{d}z\right| \le 2\int_0^{\pi/2} e^{-Ra\sin\theta}|f(Re^{i\theta})|R\,\mathrm{d}\theta \tag{1.178}$$

$$= o\left(R\int_0^{\pi/2} e^{-\frac{2aR}{\pi}\theta}\,\mathrm{d}\theta\right) = o(1-e^{-aR}) = o(1).$$

If the curve is also in the second quadrant $0 \le \theta \le \phi$, $\frac{\pi}{2} \le \phi \le \pi$, the divide the integral into two $\int_0^{\pi/2} + \int_{\pi/2}^{\phi}$ and note that the second integral is in absolute value $\int_{\pi/2}^{\pi} e^{-Ra\sin\theta}|f(Re^{i\theta})|R\,\mathrm{d}\theta$. By the change of variable $t = \pi - \theta$, $t : \frac{\pi}{2} \to 0$, we see that the latter integral becomes the same as one in (1.179).
(ii) When C_2 is the left semi-circle,

$$\left|\int_{C_2} e^{az} f(z)\,\mathrm{d}z\right| \le \int_{\pi/2}^{\pi} e^{Ra\cos\theta}|f(Re^{i\theta})|R\,\mathrm{d}\theta + \int_{\pi}^{3\pi/2} e^{Ra\cos\theta}|f(Re^{i\theta})|R\,\mathrm{d}\theta \tag{1.179}$$

$$\le 2\int_0^{\pi/2} e^{-Ra\sin\theta} o(1) R\,\mathrm{d}\theta$$

by the change of variable $t = \frac{3}{2}\pi - \theta$ for that part in the third quadrant. As in (1.179), the last integral is $o(1)$. For the arc $\widehat{AB}$ extending to the right half plane, where $B = Ri$ is the upper intercept of the circle with the imaginary axis, let the angle $\angle AOB = \alpha$ we note that in view of

$\tan\alpha \le \frac{c}{R}\left(1+O(R^{-2})\right)$. Hence

$$\left|\int_{\widehat{AB}} e^{az} f(z)\, \mathrm{d}z\right| = o(\Lambda(\widehat{AB})) = o(1),$$

where Λ indicates the length and the arc length is $\le 2\pi R \frac{c}{R} = 2\pi c$. □

Example 1.30. Let $a > 0$. We find the value of the integral

$$\int_0^\infty \frac{\sin ax}{x}\, \mathrm{d}x = \frac{\pi}{2} \tag{1.180}$$

which is the sine transform

$$\mathcal{S}[f](a) = \frac{\pi}{2}$$

with $f(t) = \frac{1}{t}$, $t > 0$. The evaluation of the special case $a = 1$ has been done in Example 1.13. It amounts to integrating the function $f(z) = \frac{e^{iz}}{z}$ around the contour C consisting of the upper semi-circle C_R of radius R going to infinity, the line segment $[-R, -r]$, the *lower* semi-circle γ_0 of radius r going to 0 and the line segment $[r, R]$. Use is made of the *Cauchy residue theorem.*

Proof. We integrate the function $f(z) = \frac{e^{iaz}}{z}$ around the contour C' consisting of the upper semi-circle C_R of radius R going to infinity, the line segment $[-R, -r]$, the *upper* semi-circle γ_0' of radius $r \to 0+$ and the line segment $[r, R]$, then by the *Cauchy integral theorem*, we have

$$\int_C f(z)\, \mathrm{d}z = 0. \tag{1.181}$$

On the other hand,

$$\begin{aligned}\int_{C'} f(z)\, \mathrm{d}z &= \int_{C_R} f(z)\, \mathrm{d}z + \int_{-R}^{-r} f(x)\, \mathrm{d}x + \int_r^R f(x)\, \mathrm{d}x \\ &+ \int_{\gamma_a'} f(z)\, \mathrm{d}z \to 2\pi i \frac{1}{2} \operatorname*{Res}_{z=0} f(z) + \mathrm{PV}\int_{-\infty}^{\infty} f(x)\, \mathrm{d}x\end{aligned} \tag{1.182}$$

as $R \to \infty$, $r \to 0+$ by Lemma 1.2 and Jordan's lemma.

Example 1.31. ([Derrick (1984), p. 171] corrected) Suppose $\alpha \ge \beta \ge 0$, $a, b \in \mathbb{R}$, and $a \ne b$. We evaluate the integral

$$\int_{-\infty}^{\infty} \frac{\sin\alpha(x-a)}{x-a} \frac{\sin\beta(x-b)}{x-b} \mathrm{d}x = \frac{\pi}{2(a-b)}(\sin\beta(a-b) - \sin\alpha(a-b)). \tag{1.183}$$

Putting $A = \alpha(z-a)$, $B = \beta(z-b)$ and noting that

$$\sin A \sin B = -\frac{1}{2}(\cos(A+B) - \cos(A-B)),$$

we integrate the function $f(z) = -\frac{1}{2}(f_1(z) - f_2(z))$ around the contour C consisting of the upper semi-circle C_R of radius R going to infinity, the line segment $[-R, -r]$, two lower semi-circles γ_a and γ_b, with center at $z = a$, $z = b$, respectively, of radius r going to 0 and the line segment $[r, R]$, where

$$\begin{aligned} f_1(z) &= \frac{e^{i(A+B)} - e^{i(A-B)}}{(z-a)(z-b)} = \frac{e^{iA}}{(z-a)} \frac{2i \sin B}{(z-b)} \\ f_2(z) &= \frac{e^{i(A-B)} - e^{-i(A+B)}}{(z-a)(z-b)} = \frac{e^{-iB}}{(z-b)} \frac{2i \sin A}{(z-a)}. \end{aligned} \tag{1.184}$$

Since

$$f(z) = -\frac{2i}{2} \frac{e^{iA} \sin B - e^{-iB} \sin A}{(z-a)(z-b)},$$

it follows that for $x \in \mathbb{R}$

$$\operatorname{Re} f(x) = \frac{2 \sin \alpha(x-a) \sin \beta(x-b)}{(x-a)(x-b)}. \tag{1.185}$$

Since

$$\begin{aligned} \operatorname*{Res}_{z=a} f_1(z) &= \frac{2i \sin \beta(a-b)}{(a-b)} \\ \operatorname*{Res}_{z=b} f_2(z) &= \frac{2i \sin \alpha(a-b)}{(a-b)}, \end{aligned} \tag{1.186}$$

we have, by the residue theorem,

$$\begin{aligned} \int_C f(z)\,\mathrm{d}z &= 2\pi i \sum_{z=a,b} \operatorname{Res} f(z) \\ &= -\pi i \operatorname*{Res}_{z=a} f_1(z) + \pi i \operatorname*{Res}_{z=b} f_2(z) \\ &= 2\frac{\pi}{a-b}(\sin \beta(a-b) - \sin \alpha(a-b)). \end{aligned} \tag{1.187}$$

On the other hand if $a < b$ and $2r \leq b - a$,

$$\begin{aligned} \int_C f(z)\,\mathrm{d}z &= \int_{C_R} f(z)\,\mathrm{d}z + \int_{-R}^{a-r} f(x)\,\mathrm{d}x + \int_{a+r}^{b-r} f(x)\,\mathrm{d}x + \int_{b+r}^{R} f(x)\,\mathrm{d}x \\ &+ \int_{\gamma_a} f(z)\,\mathrm{d}z + \int_{\gamma_b} f(z)\,\mathrm{d}z \\ &\to 2\pi i \frac{1}{2} \operatorname*{Res}_{z=a} f(z) + 2\pi i \frac{1}{2} \operatorname*{Res}_{z=b} f(z) + \mathrm{PV} \int_{-\infty}^{\infty} f(x)\,\mathrm{d}x \end{aligned} \tag{1.188}$$

as $R \to \infty$, $r \to 0+$ by Lemma 1.2. By (1.187) and (1.188), we conclude that

$$\mathrm{PV}\int_{-\infty}^{\infty} f(x)\,\mathrm{d}x = \frac{\pi}{a-b}(\sin\beta(a-b) - \sin\alpha(a-b)). \tag{1.189}$$

Taking the real part in conjunction with (1.185), we obtain

$$\mathrm{PV}\int_{-\infty}^{\infty} \frac{\sin\alpha(x-a)\sin\beta(x-b)}{(x-a)(x-b)}\,\mathrm{d}x = \frac{\pi}{2(a-b)}(\sin\beta(a-b) - \sin\alpha(a-b)). \tag{1.190}$$

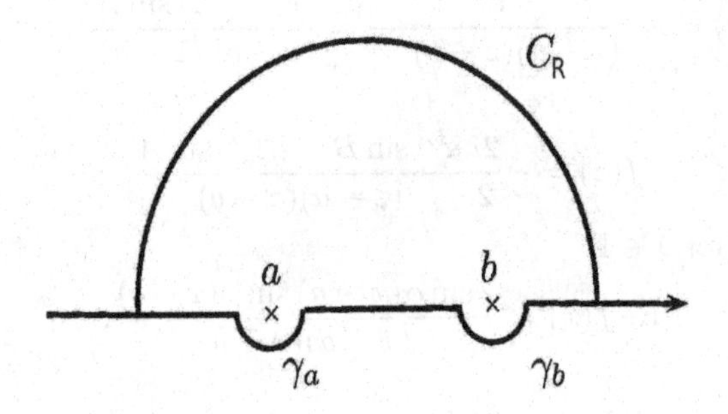

Fig. 1.9 The curve C for Example 1.31

Let as in [Woodward (1953)]

$$\mathrm{sinc}(t) = \frac{\sin\pi t}{\pi t}. \tag{1.191}$$

Theorem 1.17. *The system of sinc-functions* $\{\mathrm{sinc}(t-m),\ m \in \mathbb{Z}\}$ *form an orthonormal system (abbreviated as ONS), i.e. the orthogonality relation*

$$\int_{-\infty}^{\infty} \mathrm{sinc}(t-m)\mathrm{sinc}(t-n)\,\mathrm{d}t = \delta_{mn}, \tag{1.192}$$

holds, where δ_{mn} *indicates the Kronecker delta function and is equal to* 1 *or* 0 *according as* $m = n$ *or* $m \neq n$.

$m \neq n$ case is (1.183), so that to prove this theorem, we need one more evaluation.

Example 1.32. We have

$$\int_{-\infty}^{\infty} \frac{\sin^2 x}{x^2}\,\mathrm{d}x = \pi. \tag{1.193}$$

To see this, we note that $\sin^2 x = \frac{1-\cos 2x}{2}$, so that we integrate the function $f(z) = \frac{1-e^{2iz}}{2z^2}$ around the same contour C as in previous examples, i.e. the one consisting of the bigger upper semi-circle C_R, the line segments $[-R,-r],[r,R]$ and the lower small semi-circle γ_0 at the origin. By the same reasoning as above, we conclude that

$$\int_{-\infty}^{\infty} \frac{1-e^{2ix}}{2x^2}\,\mathrm{d}x = \pi. \tag{1.194}$$

Taking the real part proves (1.193).

Theorem 1.17 is a basis of Theorem 2.33, Shannon sampling theorem or the *sinc*-expansion.

Exercise 1.16. Confirm the following evaluations.
(i) $\int_0^\infty \frac{\log x}{x^2+1}\,\mathrm{d}x = 0$.
(ii) $\int_0^\infty \frac{\log^2 x}{x^2+1}\,\mathrm{d}x = \frac{\pi^3}{8}$.
(iii) $\int_0^\infty \frac{\log^3 x}{x^2+1}\,\mathrm{d}x = 0$.
(iv) $\int_0^\infty \frac{\log x}{(x^2+1)^2}\,\mathrm{d}x = -\frac{\pi}{4}$.
(v) $\int_0^\infty \frac{\log^2 x}{(x^2+1)^2}\,\mathrm{d}x = \frac{\pi^3}{16}$.

Example 1.33. In this example we evaluate the **probability integral** by contour integration. Let $a, b > 0$ be constants to be determined later. Let $f(z) = \frac{\exp\frac{iz^2}{2}}{e^{-bz}-1}$ and let $C = C_{R,r}$ be the contour $ABCDEFA$ consisting of the sides of the parallelogram $ADEF$, indented at the origin by the arc. More precisely, let R and r be positive numbers such that $r < R$ and $A: z = -R-iR$, $B: z = -r-ir = \sqrt{2}re^{-\frac{3}{4}\pi i}$, $C: z = r+ir = \sqrt{2}re^{\frac{1}{4}\pi i}$, $D: z = R+iR$, $E: z = R+i(R+a)$, $F: z = -R-i(R-a)$.

By Lemma 1.2, we have

$$\begin{aligned}\int_{BC} f(z)\,\mathrm{d}z &= -\int_{CB} f(z)\,\mathrm{d}z \to -\pi i \operatorname*{Res}_{z=0} f(z) \qquad (1.195)\\ &= -\pi i \lim_{z\to 0} z\frac{\exp\frac{iz^2}{2}}{e^{-bz}-1} = -\pi i \lim_{z\to 0} z\frac{\exp\frac{iz^2}{2}}{-bz+\cdots}\\ &= \frac{\pi}{b}i \quad \text{as} \quad r \to 0.\end{aligned}$$

The poles of $f(z)$ in the upper half-plane are $z = \frac{2\pi}{b}ni$, $n = 1, 2, \ldots$. We choose a, b so that no poles are contained in the parallelogram. It suffices to choose $\frac{2\pi}{b} > a$, i.e. $ab < 2\pi$. Then by the Cauchy integral theorem and (1.195), we have

$$0 = \int_{C_{R,r}} f(z)\,\mathrm{d}z \to -\frac{\pi}{b}i \quad \text{as} \quad R \to \infty,\ r \to 0. \tag{1.196}$$

Since

$$AB : z = (1+i)x, -R \le x \le -r, \quad CD : z = (1+i)x, r \le x \le R, \tag{1.197}$$

it follows that

$$\int_{CD} f(z)\,\mathrm{d}z = (1+i)\int_r^R \frac{e^{-x^2}}{B^{-1}-1}\,\mathrm{d}x, \tag{1.198}$$

where we put $B = e^{(1+i)bx}$ and that

$$\int_{AB} f(z)\,\mathrm{d}z = (1+i)\int_r^R \frac{e^{-x^2}}{B-1}\,\mathrm{d}x \tag{1.199}$$

after the change of variable $x \leftrightarrow -x$. Hence in view of the apparent equality $\frac{1}{B^{-1}-1} + \frac{1}{B-1} = -1$, we deduce that

$$\int_{AB} f(z)\,\mathrm{d}z + \int_{CD} f(z)\,\mathrm{d}z = -(1+i)\int_r^R e^{-x^2}\,\mathrm{d}x. \tag{1.200}$$

Correspondingly, we treat the integral along the side $FE : z = (1+i)x + ia, -R \le x \le R$:

$$\int_{FE} f(z)\,\mathrm{d}z = (1+i)e^{-\frac{i}{2}a^2}\int_{-R}^R \frac{e^{-x^2}A^{-1}}{e^{-abi}B^{-1}-1}\,\mathrm{d}x, \tag{1.201}$$

where we put $A = e^{(1+i)ax}$. Dividing the range of integration into $[-R, 0] \cup [0, R]$ and making the change of variable in the latter integral, we obtain

$$\int_{FE} f(z)\,\mathrm{d}z = (1+i)e^{-\frac{i}{2}a^2}\int_0^R e^{-x^2}\left(\frac{A^{-1}}{e^{-abi}B^{-1}-1} + \frac{A}{e^{-abi}B-1}\right)\mathrm{d}x. \tag{1.202}$$

For the second factor of the integrand to be transformed further, we assume that

$$ab = \pi \tag{1.203}$$

$ab = \pi$ (other choices are excluded by the condition $ab < 2\pi$). Then the second factor becomes

$$-\left(\frac{A^{-1}}{B^{-1}+1} + \frac{A}{B+1}\right) = -\left(\frac{A^{-1}B + A}{B+1}\right).$$

For this to be compressible, we would choose $A^{-1}B = 1$ or $a = b$. Then this choice necessitates the choice of values of a, b and

$$a = b = \sqrt{\pi}. \tag{1.204}$$

By (1.204), (1.202) becomes

$$\int_{EF} f(z)\,\mathrm{d}z = -i(1+i)\int_0^R e^{-x^2}\,\mathrm{d}x. \tag{1.205}$$

Incorporating the limit form $(r \to +0, R \to \infty)$ of the data (1.200) and (1.205) in (1.196), we conclude that

$$-(1+i+i(1+i))\int_0^\infty e^{-x^2}\,\mathrm{d}x = -\frac{\pi}{b}i = -\sqrt{\pi}i, \tag{1.206}$$

whence

$$\int_0^\infty e^{-x^2}\,\mathrm{d}x = \frac{\sqrt{\pi}}{2}. \tag{1.207}$$

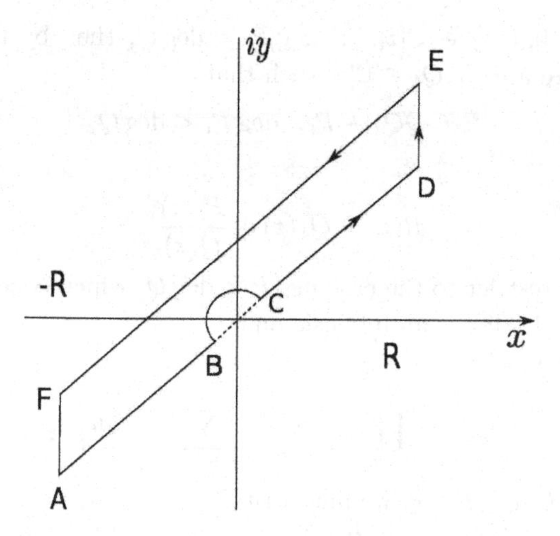

Fig. 1.10 The contour for Example 1.33

Remark 1.9. There are many known evaluations of this important value of the probability integral. Cf. Example 2.4 and Exercise A.6. But all of the evaluations based on complex integration are indirect in the sense that they give the value of a related constant which then is used to derive (1.207). The direct proof depending only by the contour integration was given by S. Rao and is compilated in [Mitrinović and Kečkić (1984)]. We note that (1.203) is a reminiscent of the modular relation due to Ramanujan which connects the values of Lambert series with parameters a and b respectively.

1.10 Partial fraction expansion

In this section we first present the partial fraction expansion of the rational functions. Then we go on to the partial fraction expansion for the cotangent function, which is a very typical and special case of the Mittag-Leffler expansion. Here we apply the Cauchy residue theorem to deduce it, which leads to a somewhat more general partial expansion formula as presented by [Titchmarsh (1939)].

1.10.1 *Partial fraction expansions for rational functions*

Let $\mathbb{C}[z]$ and $\mathbb{C}(z)$ denote the ring of polynomials and the field of rational functions with complex coefficients, respectively. Let

$$R(z) = \frac{P(z)}{Q(z)} \in \mathbb{C}(z), \tag{1.208}$$

where $Q(z) \neq 0, P, Q \in \mathbb{C}[z]$. If $\deg P \geq \deg Q$, then by the Euclidean algorithm, there are $P_1, Q_1 \in \mathbb{C}[z]$ such that

$$P = QQ_1 + P_1, \quad \deg P_1 < \deg Q,$$

whence

$$R(z) = Q_1(z) + \frac{P_1(z)}{Q(z)}.$$

Hence we may restrict to the case $\deg P < \deg Q$, which condition we may incorporate in (1.208) as an intrinsic one.

Suppose that

$$Q(z) = c_0 \prod_{k=1}^{q} (z - \beta_k)^{r_k}, \quad \sum_{k=1}^{q} r_k = \deg Q. \tag{1.209}$$

Then for each k, $1 \leq k \leq q$, we may write

$$R(z) = \frac{S_k(z)}{(z - \beta_k)^{r_k}}, \quad S_k(z) \in \mathbb{C}(z),$$

and $S_k(z)$ has no pole at $z = \beta_k$. We write

$$((z - \beta_k)^{r_k} R(z) =) \, S_k(z) = \sum_{j=0}^{r_k - 1} a_{k, r_k - j} (z - \beta_k)^j + (z - \beta_k)^{r_k} H_k(z),$$

where $H_k(z) \in \mathbb{C}(z)$ has no pole at $z = \beta_k$. By successively differentiating and setting $z = \beta_k$, we obtain

$$a_{k, r_j - j} = \frac{1}{j!} \lim_{z \to \beta_k} \frac{\mathrm{d}^j}{\mathrm{d}z^j} \left((z - \beta_k)^{r_k} R(z) \right). \tag{1.210}$$

Now, the rational function

$$F(z) := R(z) - \sum_{k=1}^{q} \sum_{j=0}^{r_k-1} a_{k,r_k-j} \frac{1}{(z-\beta_k)^{r_k-j}}$$

has no pole, so that it must be a polynomial. But, since $\lim_{z\to\infty} F(z) = 0$ (where we use the assumption $\deg P < \deg Q$), it follows that $F(z)$ must be zero.

Theorem 1.18. *If the denominator $Q(z)$ of the rational function $R(z) = \frac{P(z)}{Q(z)}$ is given by* (1.209), *then*

$$R(z) = \sum_{k=1}^{q} \sum_{j=0}^{r_k-1} a_{k,r_k-j} \frac{1}{(z-\beta_k)^{r_k-j}},$$

where the coefficients are given by (1.210).

1.11 Second-order systems and the Laplace transform

1.11.1 *Examples of second-order systems*

We give two typical examples of a second-order system.

- Electrical circuits
  The electric current $i = i(t)$ flowing an electrical circuit which consists of four ingredients, electromotive-force $e = e(t)$, resistance R, coil L and condenser C satisfies

$$L\frac{\mathrm{d}^2 i}{\mathrm{d}t^2} + R\frac{\mathrm{d}i}{\mathrm{d}t} + \frac{1}{C}i = e'(t). \tag{1.211}$$

  For when the electric current $i = i(t)$ flows the last three terminals, the inverse electro-motive force exerted on them is given, respectively by

$$e_R = -Ri, \quad e_L = -L\frac{\mathrm{d}i}{\mathrm{d}t}, \quad e_C = -\frac{1}{C}\int_0^t i\,\mathrm{d}t.$$

  The first law of Kirchhoff asserts that the sum of all inverse electromotive forces of the circuit connecting the above three terminals is 0:

$$-Ri - e_C - \frac{1}{C}\int_0^t i\,\mathrm{d}t - L\frac{\mathrm{d}i}{\mathrm{d}t} + e(t) = 0, \tag{1.212}$$

  or after differentiation, it leads to (1.211).

- Newton's equation of motion (cf. [Grodins (1963)])

$$M\frac{d^2y}{dt^2} + R\frac{dy}{dt} + Ky = e(t) = F, \tag{1.213}$$

where M is the inertance of mass, R is the viscous resistance of the dashpot and K is the spring stiffness.
Introducing the new parameters

$$\omega_n = \sqrt{\frac{K}{M}} : \text{natural angular frequence}$$

$$\zeta = \frac{R}{\sqrt{2\frac{K}{M}}} : \text{damping ratio,}$$

(1.213) becomes

$$\frac{1}{\omega_n^2}\frac{d^2y}{dt^2} + \frac{2\zeta}{\omega_n}\frac{dy}{dt} + y = \frac{1}{K}F. \tag{1.214}$$

1.11.2 *The Laplace transform method*

To solve (1.211), we use the Laplace transform method.

Definition 1.6. For functions to be Laplace-transformed, we consider only "forgetting the past functions" to the effect that the function $f(t)$ is 0 for $t < 0$, called **causal functions**. Let $u = u(t)$ be the **unit step function** $u(t) = 0, t < 0$, $u(t) = 1, t \geq 0$. By multiplying this all the functions may be thought of as causal functions. Suppose $y(t) = O(e^{at})$, $t \to \infty$ for an $a \in \mathbb{R}$. The **Laplace transform** $Y(s) = \mathcal{L}[y](s)$ of $y = y(t)$ is defined by

$$Y(s) = \mathcal{L}[y](s) = \int_0^\infty e^{-st}y(t)\,dt, \quad \operatorname{Re} s > a. \tag{1.215}$$

The integral converges absolutely in $\operatorname{Re} s > a$ and by the Weierstrass M-test, Theorem 2.8, it represents an analytic function there. Thus the domain of definition of a Laplace transform is a half-plane in the first instance, which is sometimes expressed as $\mathcal{RHP}$. Cf. §1.12 in which it means $\operatorname{Re} s > 0$, the most typical domain appearing in control theory. The Laplace transform is a (convenient form of the) Fourier transform developed in §2.13.

Corollary 1.7. (Inverse Laplace transform) *Suppose $F(s)$ is analytic except for finitely many poles in $\mathbb{C}$ and that $F(s) = o(1)$, $|s| \to \infty$. Let $c > 0$. Suppose on $\mathrm{Re}\, s = c$ there is no pole of F and let $s_1, \ldots, s_n$ be the poles of F on the left of $\mathrm{Re}\, s = c$. Then for the Laplace transform $F(s) = \mathcal{L}[f](s)$ any $t > 0$*

$$f(t) = \mathcal{L}^{-1}[F](t) = \frac{1}{2\pi i} PV \int_{Br} e^{st} F(s)\,\mathrm{d}s = \sum_{k=1}^{n} \operatorname*{Res}_{s=s_k} e^{st} F(s), \tag{1.216}$$

where more precisely, the left-hand side is $\frac{1}{2}(f(t+0) + f(t-0))$ and PV resp. Br means the **Cauchy principal value** *and the vertical path $c - i\infty \to c + i\infty$, called the* **Bromwich path** *complied in*

$$PV \int_{Br} e^{st} F(s)\,\mathrm{d}s = \lim_{R\to\infty} \int_{c-iR}^{c+iR} e^{st} F(s)\,\mathrm{d}s. \tag{1.217}$$

This is a corollary to Jordan's lemma, Lemma 1.3.

Fig. 1.11 Pierre Simon de Laplace (1749-1827)

Example 1.34. Let $\omega \in \mathbb{C}$. Then

$$\mathcal{L}[e^{\omega t}](s) = \frac{1}{s - \omega}, \tag{1.218}$$

valid for $\mathrm{Re}\, s > \mathrm{Re}\, \omega$ in the first instance. The right-hand side of (1.218) gives a meromorphic continuation of the left-hand side to the domain $\mathbb{C}\backslash\{\omega\}$. Furthermore, (1.218) also reads

$$\mathcal{L}[\sin \omega t](s) = \frac{\omega}{s^2 + \omega^2} \tag{1.219}$$

and

$$\mathcal{L}[\cos \omega t](s) = \frac{s}{s^2 + \omega^2}. \quad (1.220)$$

For inverting the Laplace transform we also need

$$\mathcal{L}[te^{\omega t}](s) = \frac{1}{(s - \omega)^2}. \quad (1.221)$$

Solution. By definition, (1.218) clearly holds true in the mentioned range. Since the right-hand side is analytic in $\mathbb{C} \setminus \{\omega\}$, the consistency theorem establishes the second assertion. (1.20) establishes (1.219) and (1.220).

A general procedure of solving a DE by the Laplace transform method consists in finding the inverse Laplace transform such as (1.218) or (1.221). The reader can see many examples where the Laplace transform is a rational function below. Since the signals are usually exponential functions, this case will cover many practical applications.

Example 1.35. We may recover (1.219) by Corollary 1.7 and residue calculus. $F(s) = \frac{\omega}{s^2+\omega^2}$ satisfies the assumptions and has simple poles at $s = i\omega, -i\omega$. Hence

$$\mathcal{L}^{-1}[F](t) = \sum \operatorname*{Res}_{s=i\omega,-i\omega} e^{st}F(s) = \frac{1}{2i}\left(e^{i\omega t} - e^{-i\omega t}\right) = \sin \omega t.$$

Exercise 1.17. Show that the solution of the autonomous DE

$$y' = \lambda y \quad (1.222)$$

is given by

$$y = y(t) = e^{\lambda t} y(0). \quad (1.223)$$

Solution. We have

$$s\mathcal{L}[y](s) - y(0) = \lambda \mathcal{L}[y](s),$$

whence $\mathcal{L}[y](s) = y(0)\frac{1}{s-\lambda}$. This can be solved as (1.223).

Example 1.36. Suppose $b > 0, c \in \mathbb{R}$ satisfy $b^2 - 4c < 0$. Then we find the current $y = y(t)$ described by the DE

$$y'' + by' + cy = e^{-\frac{b}{2}t} \sin \frac{\sqrt{4c - b^2}}{2} t \quad (1.224)$$

where the initial values are assumed to be 0: $y(0) = 0$, $y'(0) = 0$ is

$$y(t) = e^{-\frac{b}{2}t}\left(-\frac{1}{\sqrt{4c - b^2}} t \cos \frac{\sqrt{4c - b^2}}{2} t + \frac{2}{4c - b^2} \sin \frac{\sqrt{4c - b^2}}{2} t\right). \quad (1.225)$$

Proof. Let $Y(s) = \mathcal{L}[y](s)$ be the Laplace transform of $y(t)$. Then we have

$$\frac{2}{\sqrt{4c-b^2}}Y(s) = \frac{1}{(s^2+bs+c)^2}.$$

Solving the equation $s^2+bs+c=0$, we get the solutions $\alpha, \bar{\alpha}$, where $\alpha = \frac{-b+\sqrt{4c-b^2}i}{2}$, with bar denoting the complex conjugate. And we are to find the partial fraction expansion

$$\frac{1}{(s^2+bs+c)^2} = \frac{1}{(s-\alpha)^2(s-\bar{\alpha})^2} = \frac{A}{(s-\alpha)^2} + \frac{B}{s-\alpha} + \frac{\bar{A}}{(s-\bar{\alpha})^2} + \frac{\bar{B}}{s-\bar{\alpha}}. \tag{1.226}$$

The coefficients can be found by residue calculus as follows, thereby we take into account the following properties: $\alpha^2+b\alpha+c=0$, $\alpha-\bar{\alpha} = 2i\,\mathrm{Im}\,\alpha = \sqrt{4c-b^2}i$. To find A, B we clear the denominators by multiplying both sides of (1.226) by $(s-\alpha)^2$ to find

$$\begin{aligned} \frac{1}{(s-\bar{\alpha})^2} &= \frac{(s-\alpha)^2}{(s^2+bs+c)^2} \qquad (1.227)\\ &= A + B(s-\alpha) + (s-\alpha)^2\left(\frac{\bar{A}}{(s-\bar{\alpha})^2} + \frac{\bar{B}}{(s-\bar{\alpha})}\right), \end{aligned}$$

the right side being the Taylor expansion of the left-hand side. Letting $s \to \alpha$, we find the y-intercept A: $A = \lim_{s\to\alpha}\frac{1}{(s-\bar{\alpha})^2} = \frac{1}{(\alpha-\bar{\alpha})^2} = -\frac{1}{4c-b^2}$. To determine B we differentiate both sides of (1.227) to find that

$$\begin{aligned} \frac{-2}{(s-\bar{\alpha})^3} &= \frac{\mathrm{d}}{\mathrm{d}s}\frac{1}{(s-\bar{\alpha})^2} \qquad (1.228)\\ &= B + (s-\alpha)(\cdots), \end{aligned}$$

whence $B = \lim_{s\to\alpha}\frac{-2}{(s-\bar{\alpha})^3} = \frac{-2}{(\alpha-\bar{\alpha})^3} = -\frac{2\sqrt{4c-b^2}}{(4c-b^2)^2}i$. Hence we conclude that

$$\frac{1}{(s^2+bs+c)^2} = \frac{-\frac{1}{4c-b^2}}{(s-\alpha)^2} + \frac{-\frac{2\sqrt{4c-b^2}}{(4c-b^2)^2}i}{s-\alpha} + \frac{-\frac{1}{4c-b^2}}{(s-\bar{\alpha})^2} + \frac{\frac{2\sqrt{4c-b^2}}{(4c-b^2)^2}i}{s-\bar{\alpha}}. \tag{1.229}$$

We have therefore

$$\begin{aligned} &\frac{2}{\sqrt{4c-b^2}}y(t) \qquad (1.230)\\ &= \frac{2}{\sqrt{4c-b^2}}\mathcal{L}^{-1}[\mathcal{L}[y]](t)\\ &= -\frac{2}{4c-b^2}te^{-\frac{b}{2}t}\cos\frac{\sqrt{4c-b^2}}{2}t + \frac{4}{(4c-b^2)^{3/2}}e^{-\frac{b}{2}t}\sin\frac{\sqrt{4c-b^2}}{2}t, \end{aligned}$$

or (1.225), completing the proof.

Remark 1.10. For (1.230), cf. Remark 1.8. Although the DE (1.224) may be considered for any b, the reason why we restricted b to be positive can be seen from (1.225) in which the current goes to 0 only when $b > 0$. This case is known as a **transient state** which is the state of a circuit without current in the beginning $y(0) = 0$ and exerted current in one instant $e(t) = e^{-\frac{b}{2}t} \sin \frac{\sqrt{4c-b^2}}{2}t$ and immediate shut-down of switch. Then the exponential decrease of the current in the circuit obeys the law (1.225).

Although the following exercise are special cases of Example 1.36, the readers are advised to solve them independently.

Exercise 1.18. Find the current $y = y(t)$ described by the DE

$$y'' + y = \sin t,$$

where the initial values are assumed to be 0: $y(0) = 0$, $y'(0) = 0$.

Solution.

$$y(t) = \frac{1}{2}(-t\cos t + \sin t).$$

Exercise 1.19. Find the current $y = y(t)$ described by the DE

$$y'' + y' + y = e^{-\frac{1}{2}t} \sin \frac{\sqrt{3}}{2}t,$$

where the initial values are assumed to be 0: $y(0) = 0$, $y'(0) = 0$.

Solution. Let $\mathcal{L}[y](s)$ be the Laplace transform of $y(t)$. Then we have

$$\frac{2}{\sqrt{3}}\mathcal{L}[y](s) = \frac{1}{(s^2+s+1)^2}$$

and we may obtain the partial fraction expansion

$$\frac{1}{(s^2+s+1)^2} = \frac{-\frac{1}{3}}{(s-\rho)^2} + \frac{-\frac{2}{3\sqrt{3}}i}{s-\rho} + \frac{-\frac{1}{3}}{(s-\bar{\rho})^2} + \frac{\frac{2}{3\sqrt{3}}i}{s-\bar{\rho}},$$

where $\rho = e^{\frac{2\pi i}{3}} = \frac{-1+\sqrt{3}i}{2}$ is the first primitive cube root of 1. Hence

$$\frac{2}{\sqrt{3}}y(t) = \frac{2}{\sqrt{3}}\mathcal{L}^{-1}[\mathcal{L}[y]](t) = -\frac{2}{3}te^{-\frac{1}{2}t}\cos\frac{\sqrt{3}}{2}t + \frac{4}{3\sqrt{3}}e^{-\frac{1}{2}t}\sin\frac{\sqrt{3}}{2}t.$$

Exercise 1.20. Find the solution of the DE

$$y'' - y' + y = e^{\frac{1}{2}t} \sin \frac{\sqrt{3}}{2}t,$$

where the initial values are assumed to be 0: $y(0) = 0$, $y'(0) = 0$.

Solution. We have

$$\frac{2}{\sqrt{3}}\mathcal{L}[y](s) = \frac{1}{(s^2 - s + 1)^2}$$

and

$$\frac{1}{(s^2 - s + 1)^2} = \frac{-\frac{1}{3}}{(s-\alpha)^2} + \frac{-\frac{2}{3\sqrt{3}}i}{s-\alpha} + \frac{-\frac{1}{3}}{(s-\bar{\alpha})^2} + \frac{\frac{2}{3\sqrt{3}}i}{s-\bar{\alpha}}.$$

Hence

$$\frac{2}{\sqrt{3}}y(t) = \frac{2}{\sqrt{3}}L^{-1}[L[y]](t) = -\frac{2}{3}te^{\frac{1}{2}t}\cos\frac{\sqrt{3}}{2}t + \frac{4}{3\sqrt{3}}e^{\frac{1}{2}t}\sin\frac{\sqrt{3}}{2}t.$$

1.11.3 *Partial fraction expansion for the cotangent function and some of its applications*

The contents of this subsection has much relevance to those of §2.10.

Theorem 1.19. (Partial fraction expansion for cot) *For non-integral values of z, we have the partial fraction expansion*

$$\cot \pi z = \frac{1}{\pi z} + \frac{1}{\pi}\sum_{n=1}^{\infty}\left(\frac{1}{z-n} + \frac{1}{z+n}\right) \tag{1.231}$$
$$= \frac{1}{\pi z} + \frac{2z}{\pi}\sum_{n=1}^{\infty}\frac{1}{z^2 - n^2}.$$

Proof. Let C_N denote the square with center at the origin and with sides parallel to the coordinate axes, the vertices being at the points $\pm(N+\frac{1}{2}) \pm i(N+\frac{1}{2})$, where $N \in \mathbb{N}$.

Consider the integral

$$I_N = \frac{1}{2\pi i}\int_{C_N}\frac{f(w)}{w(w-z)}dw, \tag{1.232}$$

where

$$f(w) = \cot \pi w - \frac{1}{\pi w}.$$

By L'Hospital's rule, we see that

$$\lim_{w \to 0} f(w) = 0,$$

and so the origin is a removable singularity of the integrand. Hence the (simple) poles of the integrand are at $w = k, -N \le k \le N, k \neq 0$ and $w = z$, so that by the Cauchy residue theorem, we have

$$I_N = f(z) + \sum_{\substack{k=-N \\ k \neq 0}}^{N} \frac{\frac{1}{\pi}}{k(k-z)}, \tag{1.233}$$

where we used the fact that

$$\operatorname*{Res}_{w=k} f(w) = \frac{1}{\pi},$$

which is immediately seen from the limit relation

$$\lim_{w\to k}(w-k)\cot\pi w = \frac{1}{\pi}\lim_{w\to 0}\frac{\pi(w-k)}{\sin\pi(w-k)}\cos\pi(w-k) = \frac{1}{\pi}.$$

If we can show that $I_N \to 0$ as $N\to\infty$, then (1.233) shows that the series $\sum_{k=-\infty,k\neq 0}^{\infty}\frac{\frac{1}{\pi}}{k(k-z)}$ is convergent and (1.231) follows. The convergence of the series is apparent because it is comparable to $\sum_{n=1}^{\infty}\frac{1}{n^2}$. Therefore it is uniformly convergent in the domain without integer points.

We have the estimates.

On the vertical sides $w = \pm(N+\frac{1}{2}) + iy, -N < y < N$, we have

$$\left|\cot\pi\left(iy\pm\left(N+\frac{1}{2}\right)\right)\right| = |\tan\pi iy| = \frac{e^{\pi y}-e^{-\pi y}}{e^{\pi y}+e^{-\pi y}} < 1,$$

while on the horizontal sides $w = x\pm\left(N+\frac{1}{2}\right)i, -N < x < N$, we have

$$\left|\cot\pi\left(x\pm\left(N+\frac{1}{2}\right)i\right)\right| < \frac{e^{\pi(N+\frac{1}{2})}+e^{-\pi(N+\frac{1}{2})}}{e^{\pi(N+\frac{1}{2})}-e^{-\pi(N+\frac{1}{2})}} = \frac{1+e^{-2\pi(N+\frac{1}{2})}}{1-e^{-2\pi(N+\frac{1}{2})}} < 2,$$

for $N > N_0$, where N_0 is a constant. Hence $|f(w)| < 2$ on C_N. Since the length of $C_N = 4\left(N+\frac{1}{2}\right)$ and the denominator $w(w-z)\sim N^2$, it follows that

$$|I_N| < \frac{5}{N},$$

whence $\lim_{N\to\infty} I_N = 0$, completing the proof of our theorem. □

We note that the above procedure is a refinement of [Gel'fond (1971), p. 33] and it can be immediately generalized into a theorem in [Titchmarsh (1939), pp. 110].

Theorem 1.20. (Titchmarsh) *Suppose that $f(z)$ is a meromorphic function with simple poles at a_k, $k\in N$ with $0<|a_1|\le|a_2|\le\cdots$ with residue $b_k = \operatorname{Res}_{z=a_k} f(z)$. Suppose there is a sequence of closed contours C_N, $N\in\mathbb{N}$, including $a_1,\ldots,a_N$ only, such that the minimum distance R_N of C_N from the origin tends to ∞ as $N\to\infty$ while the circumference L_N of C_N is $O(R_N)$ and such that $f(z) = o(R_N)$ on C_N. Under these conditions*

$$f(z) = f(0) + \sum_{n=1}^{\infty} b_n\left(\frac{1}{z-a_n}+\frac{1}{a_n}\right) \tag{1.234}$$

for all values of z except for poles.

Example 1.37. We deduce the partial fraction expansion for the cosecant function

$$\pi \operatorname{cosec} \pi z - \frac{1}{z} = 2\sum_{k=1}^{\infty} \frac{(-1)^k z}{z^2 - k^2} \tag{1.235}$$

as a consequence of Theorem 1.20. Let C_N denote the same contour as in Theorem 1.19, and

$$f(w) = \frac{1}{\sin \pi w} - \frac{1}{\pi w}.$$

Then $w = 0$ is a removable singularity and $f(w)$ has simple poles at $a_k = k, 0 \neq k \in \mathbb{Z}$ with residue

$$b_k = \frac{1}{\pi} \lim_{w \to k} \pi(w-k) \frac{1}{\sin \pi w} = \frac{(-1)^k}{\pi} \lim_{w \to k} \frac{\pi(w-k)}{\sin \pi(w-k)} = \frac{(-1)^k}{\pi}.$$

Hence Theorem 1.20 implies that

$$f(z) = \sum_{\substack{k=-\infty \\ k \neq 0}}^{\infty} \frac{(-1)^k}{\pi} \left(\frac{1}{z-k} + \frac{1}{k} \right),$$

so that

$$\operatorname{cosec} \pi z - \frac{1}{\pi z} = \sum_{k=1}^{\infty} \frac{(-1)^k}{\pi} \left(\frac{1}{z-k} + \frac{1}{z+k} \right) \tag{1.236}$$

on coupling the terms with k and $-k$, which we may because the series for $k \in \mathbb{N}$ and $-k$ both converge. (1.236) leads to (1.235).

Now, as an immediate corollary to (1.231), we deduce the product representation for the sine function.

Theorem 1.21. *For all z we have the product representation*

$$\frac{\sin \pi z}{\pi z} = \prod_{n=1}^{\infty} \left(1 - \frac{z^2}{n^2} \right), \tag{1.237}$$

the product being absolutely convergent.

Proof. Integrating (1.231) multiplied by π from ζ and z, we obtain

$$\log \sin \pi z - \log \sin \pi \zeta = \int_{\zeta}^{z} \pi \cot \pi z dz = \log \frac{z}{\zeta} + \sum_{n=1}^{\infty} \log \frac{z-n}{\zeta-n} \frac{z+n}{\zeta+n}.$$

After simplification, we have

$$\log \sin \pi z = \log \sin \pi \zeta \cdot \frac{z \prod_{n=1}^{\infty}\left(1 - \frac{z^2}{n^2}\right)}{\zeta \prod_{n=1}^{\infty}\left(1 - \frac{\zeta^2}{n^2}\right)},$$

whence

$$\sin \pi z = \pi z \prod_{n=1}^{\infty}\left(1 - \frac{z^2}{n^2}\right)\left(\frac{\pi\zeta \prod_{n=1}^{\infty}\left(1 - \frac{\zeta^2}{n^2}\right)}{\sin \pi\zeta}\right)^{-1}.$$

Passing to the limit as $\zeta \to 0$, we get (1.237). □

Remark 1.11. (i) In [Kanemitsu and Tsukada (2007)], (1.237) is deduced from the Weierstrass product for the gamma function and the reciprocal relation between the gamma and sine functions.
(ii) (1.237) is of historical interest.

By (1.14), the left-hand side has the Maclaurin expansion (cf. (29.1))

$$\frac{\sin \pi z}{\pi z} = 1 - \frac{1}{3!}(\pi z)^2 + \frac{1}{5!}(\pi z)^4 - \cdots, \tag{1.238}$$

while the right-hand side becomes, after multiplying out

$$1 - \left(\sum_{n=1}^{\infty} \frac{1}{n^2}\right) z^2 + \left(\sum_{m,n=1}^{\infty} \frac{1}{(mn)^2}\right) z^4 - \cdots.$$

Hence, comparing the coefficients, we obtain

$$\zeta(2) = \sum_{n=1}^{\infty} \frac{1}{n^2} = \frac{\pi^2}{6}, \tag{1.239}$$

which is the famous solution to the Basler problem due to Euler, and

$$\sum_{m,n=1}^{\infty} \frac{1}{m^2 n^2} = \frac{\pi^4}{5!}. \tag{1.240}$$

The left-hand side of (1.240) is written as

$$\sum_{m=n} + \sum_{m>n} + \sum_{m<n} = \zeta(4) + 2\sum_{m=2}^{\infty} \frac{1}{m^2} \sum_{n<m} \frac{1}{n^2},$$

which is a prototype of the **Euler-Zagier sum.**

We describe the **principle of applying the residue theorem to the cut domain with the cut along the positive real axis** by the following

Theorem 1.22. *For $a \in \mathbb{R}$, $a \notin \mathbb{Z}$, we choose the branch $0 \le \arg z^a < 2\pi$. Suppose $f(z)$ is a rational function without a pole on the real axis and satisfies the growth conditions.*

$$\lim_{z\to 0} z^{a+1} f(z) = 0, \quad \lim_{z\to\infty} z^{a+1} f(z) = 0.$$

Then we have

$$\int_0^\infty x^a f(x)\,\mathrm{d}x = \frac{2\pi i}{1-e^{2\pi i a}} \sum_{z\neq 0} \operatorname{Res} z^a f(z). \tag{1.241}$$

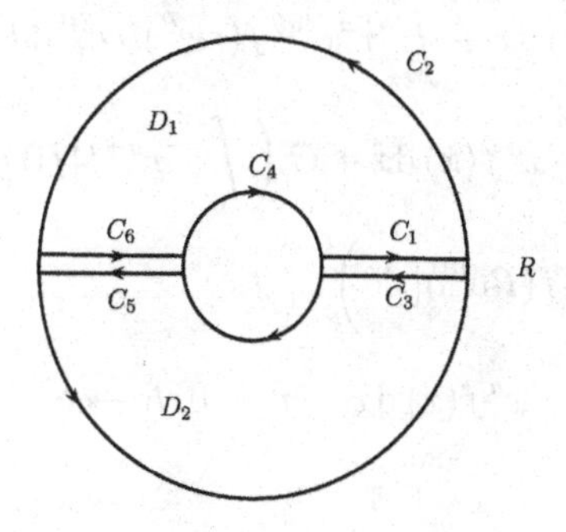

Fig. 1.12 Two simply-connected domains

Proof. Since $f(z)$ has only finitely many poles, the poles different from the origin lies in the ring-shaped domain $r < |z| < R$. We choose the branch of $f(z)$ such that $0 \le \arg z^a < 2\pi$. Consider two domains that overlap the line segment C_1, C_3 and C_5, C_6, where

$$\partial D_1 = C_1 + \text{upper half of } C_2 + C_6 + \text{upper half of } C_4,$$
$$\partial D_1 = C_3 + \text{lower half of } C_4 + C_5 + \text{lower half of } C_2$$

and

$C_1 : z = x, x : r \to R; C_2 : z = Re^{i\theta}, 0 \le \theta \le 2\pi; C_3 : z = x, x : R \to r$
$C_4 : z = re^{i\theta}, 0 \le \theta \le 2\pi; C_5 : z = xe^{\pi i}, x : r \to R; C_6 : z = xe^{\pi i}, x : R \to r.$

Hence on C_5, C_6 we have $z^a f(z) = x^a e^{\pi i a} f(-x)$ and since C_5, C_6 are curves of opposite direction, we have

$$\int_{C_5} + \int_{C_6} z^a f(z)\,\mathrm{d}z = 0.$$

Let D denote the domain $D_1 \cup D_2$ removed C_5, C_6, which is the cut domain with a cut along the positive real axis.C_1, C_2 are called the upper ridge and the lower ridge of the cut. Then the integral along the boundary $\partial D_1 + \partial D_2$ is the same as the one along $\partial D = C_1 + C_2 + C_3 + C_4$. We describe this as the above principle. Hence

$$
\begin{aligned}
&\int_{\partial D} z^a f(z)\,\mathrm{d}z = \int_{C_1} + \int_{C_2} + \int_{C_3} + \int_{C_4} \\
&= \int_r^R x^a f(x)\,\mathrm{d}x + \int_0^{2\pi} R^a e^{ia\theta} f(Re^{i\theta}) iRe^{i\theta}\,\mathrm{d}\theta \\
&+ \int_R^r x^a e^{2\pi ia} f(x)\,\mathrm{d}x + \int_{2\pi}^0 r^a e^{ia\theta} f(re^{i\theta}) ire^{i\theta}\,\mathrm{d}\theta \\
&= (1 - e^{2\pi ia}) \int_r^R x^a f(x)\,\mathrm{d}x + O\left(\int_0^{2\pi} r^{a+1} |f(re^{i\theta})|\,\mathrm{d}\theta\right) \\
&+ O\left(\int_0^{2\pi} R^{a+1} |f(Re^{i\theta})|\,\mathrm{d}\theta\right) \\
&\to (1 - e^{2\pi ia}) \int_0^\infty x^a f(x)\,\mathrm{d}x, \quad r \to 0, R \to \infty.
\end{aligned}
\tag{1.242}
$$

Since the left-hand side member of (1.242) is the sum of all the residues excpet possible at $z = 0$, we conclude (1.241). □

Remark 1.12. In the above **principle of applying the residue theorem to the cut domain**, one needs to pay attention that $f(z)$ is not single-valued in D and the residue theorem cannot be applied directly. In practice, we apply the theorem to two simply connected domains D_1, D_2 where $f(z)$ is single-valued, where the residue theorem may be applied. Then as we have seen, the integrals on the overlapping part cancel each other, which makes it look like we apply the residue theorem to the domain D.

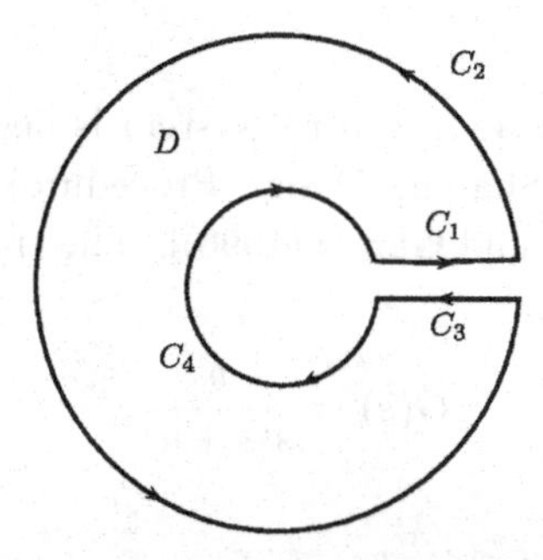

Fig. 1.13 Cut domain

Exercise 1.21. Confirm the following evaluations, where $b > 0$.
(i) $\int_0^\infty \frac{x^a}{x^2+b^2}\,dx = \frac{\pi}{2} b^{a-1} \sec \frac{\pi a}{2}$ for $-1 < a < 1$.
(ii) $\int_0^\infty \frac{x^a}{(x^2+b^2)^2}\,dx = \frac{\pi}{4} b^{a-3}(1-a) \sec \frac{\pi a}{2}$ for $-1 < a < 3$.

These are examples of the Mellin transform, cf. [Kanemitsu and Tsukada (2007), pp. 161-163]. It is a version of the Laplace transform which in turn is a version of the Fourier transform discussed in § 1.11.2 and § 2.13, respectively. For another example, cf. the gamma function discussed in Example 42, (i) and Exercise A.6.

1.12 Robust controller for servo systems

In this section we state the results of the paper of Zou [Zou] where by introducing some modifications in Kimura's argument ([Kimura (1997)]), a proof is given of a general theorem for the celebrated Nevanlinna-Pick theorem as proved in [Kimura (1984)] in terms of the J-lossless conjugation, with all parametric solutions given. A resort is made to Kimura's stabilization method ([Kimura (1984)]) for a servo system and elucidate the result in [Ban *et al.* (2009)]. It is possible to separate the essential ingredient—a solution to the generalized Nevenlinnan-Pick interpolation problem from an elementary method for determining the form of the linear fractional transformation by symmetry points. For an accessible introduction to and terminologies in control theory, we refer to [Li (2012), Chapter 6].

1.12.1 *Prerequisites*

In [Ban *et al.* (2009)], a servo control system is designed for a DC motor using the LSDP (Loop Shaping Design Procedure) method of McFarlane and Glover [MacFarlane and Glover (1996)]. The transfer function is given by

$$G(s) = \frac{b}{s(s+a)}, \tag{1.243}$$

where

$$b = \frac{K_m}{J_{eq}R_m}, \; a = K_m b = \frac{K_m^2}{J_{eq}R_m}, \tag{1.244}$$

both of which are positive constants, where J_{eq} is the moment of inertia, R_m the armature resistance, and K_m the torque of the motor.

This is exactly the case considered by [Kimura (1984)] as an extension of the robust stabilizability of a plant whose unstable poles lie in the right half-plane (1.22) to the one with integrator. In what follows we use the symbol j to mean the imaginary unit i.

Definition 1.7. A minimal phase factor $f_m(s)$ of $f(s)$ is one having no zero in $\mathrm{Re}\, s \geq 0$ and such that $|f_m(j\omega)| = |f(j\omega)|$. Let $C(p_0, r)$ denote the class of functions with nominal mode p_0 and uncertainty band function r,

$$r(s) = \frac{r'_m(s)}{s}, \tag{1.245}$$

where r'_m is the minimal phase function (where we follow the notation of [Kimura (1984)] and the prime does not mean the derivative). A transfer function is said to belong to the class $C(p_0, r)$ if (i) $p(s)$ has the same number of unstable poles as $p_0(s)$, where $p_0(s)$ is the nominal model of the plant dynamics with a simple pole at $s = 0$,
(ii) $|p(j\omega) - p_0(j\omega)| \leq |r(j\omega)|, \quad |r(j\omega)| > 0$ for $\forall \omega \in \mathbb{R}$.

Definition 1.8. Consider a unity feedback system Σ consisting of the input r, output y, the controller (compensator) C (also often denoted K) and the plant P, with intermediate input to and output from C denoted by e and u, respectively. The class $C(p_0, r)$ is said to be robustly stabilizable if there exists a controller C such that the unity feedback system Σ is stable for each $p \in C(p_0, r)$. Such a controller is called a robust stabilizer.

Theorem 1.23. ([Kimura (1984), Theorem 2]) *Suppose* (A1) *All the unstable poles of* $p_0(s)$ *are simple and lie in* $\mathcal{RHP}$ *and* $p_0(s)$ *has a simple pole* *at $s = 0$.* (A2) *The relative degree of the denominator of $r(s)$ is at most* 1.

Then the class of plants are robustly stabilizable if and only if the Pick matrix P in (1.250) *is positive definite and*

$$|\beta_0| = |r'_m(0)/\tilde{p}_0(0)| < 1, \tag{1.246}$$

where $\tilde{p}_0$ is defined by (1.290) *and is the nominal plant $p_0(s)$ all of the poles eliminated.*

1.12.2 A generalized Nevanlinna-Pick interpolation problem

We follow the notation in [Kimura (1997)]. Let H_∞ denote the space of all functions analytic in the right half-plane $\mathcal{RHP}$ with $\text{esssup}|f(j\omega)| < \infty$. BH^∞ indicates the subspace of H_∞ whose elements have the H_∞ norm < 1, i.e. $|f(j\omega)| < 1$ for all ω ([Kimura (1997), p. 42]).

The generalized Nevanlinna-Pick problem (GNP) states that given $(\alpha_i, \beta_i) \in \mathbb{C}$, $1 \le i \le \ell$ and additionally $\beta_0, \beta_{\ell+1} \in \mathbb{C}$ such that

$$\operatorname{Re}\alpha_i > 0, \quad |\beta_i| < 1, \quad 1 \le i \le \ell, \tag{1.247}$$

one is supposed to find a function $u \in BH^\infty$ satisfying the interpolation conditions

$$u(\alpha_i) = \beta_i, \quad 1 \le i \le \ell, \quad u(0) = \beta_0, \; u(\infty) = \beta_{\ell+1}. \tag{1.248}$$

Remark 1.13. The additional interpolation conditions in (1.248) need some explanation. The value at infinity pertains to the properness of the controller $c(s)$ expressed in (A1) in terms of $u(s)$ by (1.295) and $\beta_{\ell+1} = 0$ will suffice.

On the other hand, Lemma 1.8 below implies that all the unstable poles of $p_0(s)$ are to be canceled out by the zeros of $q(s)$, including the simple pole at $s = 0$, which imposes the other interpolation condition on $u(s)$, i.e.

$$|u(0)| = |\beta_0|, \tag{1.249}$$

which is the essential addition to the original NP problem.

Since the origin and the point at infinity are symmetric points (relative to the fractional transformations, cf. [Ahlfors (1979)]), the values at those points may be conveniently used to determine the linear fractional transformations (cf. Lemma 1.5 below).

The Hermitian matrix P defined by

$$P = \begin{pmatrix} \frac{1-\beta_1\bar{\beta}_1}{\alpha_1+\bar{\alpha}_1} \cdots & \frac{1-\beta_1\bar{\beta}_\ell}{\alpha_1+\bar{\alpha}_{ell}} \\ \cdots \quad \cdots & \cdots \\ \frac{1-\beta_\ell\bar{\beta}_1}{\alpha_\ell+\bar{\alpha}_1} \cdots & \frac{1-\beta_\ell\bar{\beta}_\ell}{\alpha_\ell+\bar{\alpha}_{ell}} \end{pmatrix} \tag{1.250}$$

is called the Pick matrix. Let

$$B_i(s) = \frac{\alpha_i - s}{\bar{\alpha}_i + s}, \; 1 \le i \le \ell, \tag{1.251}$$

which is a BH$^\infty$-function satisfying $|B_i(j\omega)| = 1$ for all $\omega \in \mathbb{R}$. Let $\beta_{i,k+1}$ denote the Fenyves array defined by

$$\beta_{i,1} = \beta_i, \; 1 \le i \le \ell, \tag{1.252}$$

$$\beta_{i,k+1} = \frac{(\alpha_i + \bar{\alpha}_i)(\beta_{i,k} - \beta_{k,k})}{(\alpha_i - \alpha_k)(1 - \bar{\beta}_{k,k}\beta_{i,k})}, \; 1 \le k \le i - 1 \le \ell - 1.$$

Let

$$\rho_k = \beta_{k,k}, \; 1 \le k \le \ell. \tag{1.253}$$

It is well-known that P is positive definite if and only if $|\beta_{i,k+1}| < 1$. Define

$$\Theta_k(s) = \frac{1}{\sqrt{1 - |\rho_k|^2}} \begin{pmatrix} -B_k(s) & \rho_k \\ -\bar{\rho}_k B_k(s) & 1 \end{pmatrix}, \; 1 \le k \le \ell. \tag{1.254}$$

In its classical version, the Nevanlinna-Pick theorem reads

Theorem 1.24. *The solvability of the NP problem is equivalent to the positive definiteness of the Pick matrix.*

Since

$$G(s) = \left(\begin{array}{c|c} A & B \\ \hline I & O \end{array}\right) = I(sI - A)^{-1}B + O = (sI - A)^{-1}B, \tag{1.255}$$

we say that the J-lossless matrix $\Theta = \Theta(s)$ is a J-lossless stabilizing conjugator of the pair (A, B) ([Kimura (1997), p. 113]) if it stabilizes G in the sense of [Kimura (1997), Definition 5.1, p. 108], i.e. if $\Theta(s)$ is a rational function with minimum degree which cancels all the unstable poles of $G(s)$.

The following lemma is a combination of [Kimura (1997), Theorem 5.2] and [Kimura (1997), Lemma 5.6].

Lemma 1.4. *The pair (A, B) in* (1.255) *has a J-lossless stabilizing (anti-stabilizing) conjugator $\Theta(s)$ if and only if the algebraic Riccati equation*

$$XA + {}^tAX - XBJ^tBX = O \tag{1.256}$$

has solution $X \ge 0$ such that $A - XBJ^tBX$ is stable (anti-stable).

Further, if A is stable ($-A$ is anti-stable), then the solution X of (1.256) *is invertible.*

We recall the definition of the homographic transformation [Kimura (1997), (4.82), p. 93], [Li (2012), p. 165]. On a controller S, the matrix

$$\Theta = \begin{pmatrix} \Theta_{11} & \Theta_{12} \\ \Theta_{21} & \Theta_{22} \end{pmatrix} \tag{1.257}$$

acts as

$$\Theta S = (\Theta_{11}S + \Theta_{12})(\Theta_{21}S + \Theta_{22})^{-1} =: \mathrm{HM}(\Theta, S). \tag{1.258}$$

Then we have the concatenation rule (cf. [Kuroš (1974), pp. 16-17], [Li (2012), p. 42])

$$\mathrm{HM}(\Theta_2, \mathrm{HM}(\Theta_1, S)) = \mathrm{HM}(\Theta_2\Theta_1, S). \tag{1.259}$$

In our present case, the controller is a scalar $s \in \mathbb{C}$ and (1.258) amounts to

$$\mathrm{HM}(\Theta, S) = \Theta S = \frac{\Theta_{11}s + \Theta_{12}}{\Theta_{21}s + \Theta_{22}}, \tag{1.260}$$

where $\Theta_{k\ell}$ are also scalars.

Lemma 1.5. *Suppose*

$$\Theta = \begin{pmatrix} t_{11} & t_{12} \\ t_{21} & t_{22} \end{pmatrix} \tag{1.261}$$

is in $\mathrm{GL}_2(\mathbb{C})$,

$$u_1 = \frac{t_{11}u_{\ell+1} + t_{12}}{t_{21}u_{\ell+1} + t_{22}} \tag{1.262}$$

and that the inverse transformation of (1.262) *is given by*

$$u_{\ell+1}(s) = f_1(s)\frac{u_1(s) + f_3(s)}{1 + f_2(s)u_1(s)}, \tag{1.263}$$

where

$$f_1(s) = \frac{t_{22}(s)}{t_{11}(s)},\ f_2(s) = -\frac{t_{21}(s)}{t_{11}(s)},\ f_3(s) = -\frac{t_{12}(s)}{t_{22}(s)}. \tag{1.264}$$

Let

$$\delta_0(= u_{\ell+1}(0)) = f_1(0)\frac{u_1(0) + f_3(0)}{1 + f_2(0)u_1(0)}, \tag{1.265}$$

$$\delta_\infty(= u_{\ell+1}(\infty)) = f_1(\infty)\frac{u_1(\infty) + f_3(\infty)}{1 + f_2(\infty)u_1(\infty)}$$

and suppose

$$|\delta_0| < 1, \quad |\delta_{\ell+1}| < 1. \tag{1.266}$$

Then the functions $u_{\ell+1} \in \mathrm{BH}^\infty$ *satisfying* (1.274) *are given by*

$$u_{\ell+1}(s) = \frac{cu_{\ell+1}(\infty)s + du_{\ell+1}(0)}{cs + d}. \tag{1.267}$$

Proof. By (1.266) we may choose the values of $u_{\ell+1}$ at $0, 1$ as in (1.265) to obtain

$$|u_{\ell+1}(0)| < 1, \quad |u_{\ell+1}(\infty)| < 1, \tag{1.268}$$

whence we only need to determine the form of the linear transformation $u_{\ell+1}$. This is a well-known procedure (e.g. [Ahlfors (1979)], $0, \infty$ are symmetrical points) and we readily conclude (1.267). □

Kimura gives the example with $c = d = 1$:

$$u_{\ell+1}(s) = \frac{u_{\ell+1}(\infty)s + u_{\ell+1}(0)}{s+1} = \frac{\delta_\infty s + \delta_0}{s+1}.$$

We need the following basic lemma which we will take for granted. It depends on the well-known bounded real lemma ([Kimura (1997), Theorem 4.5]) which connects boundedness of a transfer function and the algebraic Riccati equation.

Lemma 1.6. ([Kimura (1997), Theorem 4.5]) *Suppose Θ is (J, J')-unitary. Then there exists a termination u for which $\mathrm{HM}(\Theta, u) \in \mathrm{BH}^\infty$ if and only if Θ is (J, J')-lossless. If this is the case, then $\mathrm{HM}(\Theta, s) \in \mathrm{BH}^\infty$ if and only if $u \in \mathrm{BH}^\infty$.*

The following theorem is a slightly modified version of [Kimura (1997), Theorem 5.8, p. 115] in conjunction with [Kimura (1984)].

Theorem 1.25. *The GNP is solvable if and only if the pair of matrices*

$$A = \begin{pmatrix} \alpha_1 & 0 & \cdots & 0 \\ 0 & \alpha_2 & \cdots & 0 \\ & & \cdots & \\ 0 & 0 & \cdots & \alpha_\ell \end{pmatrix}, \quad B = \begin{pmatrix} 1 & -\beta_1 \\ 1 & -\beta_2 \\ & \cdots \\ 1 & -\beta_\ell \end{pmatrix} \tag{1.269}$$

has a J-lossless stabilizing conjugator Θ for the signature matrix

$$J = \begin{pmatrix} I & O \\ O & -I \end{pmatrix}, \tag{1.270}$$

and

$$|\beta_0| < 1, \quad |\beta_\ell| < 1. \tag{1.271}$$

If this is the case, then all the parametric solutions u_1 are obtained by the procedure

$$u_k = \mathrm{HM}(\Theta_k, u_{k+1}), \quad 1 \le k \le \ell \tag{1.272}$$

starting from any $u_{\ell+1} \in \mathrm{BH}^\infty$ *satisfying Condition* (1.283), *where* Θ_k *is defined by* (1.254). *I.e. if* Θ *is defined by*

$$\Theta = \prod_{k=1}^{\ell} \Theta_k, \tag{1.273}$$

then $\mathrm{HM}(\Theta, u_{\ell+1})$ *solves the NP problem with the additional interpolation conditions enunciated in* (1.248)

$$u_1(0) = \beta_0, \quad u_1(\infty) = \beta_{\ell+1}. \tag{1.274}$$

Proof. We mix the two proofs given by [Kimura (1984)], [Kimura (1997)]. Let Θ be defined by (1.261). Then by (1.272), (1.273) and the concatenation rule we have ((1.262))

$$u_1 = \mathrm{HM}(\Theta, u_{\ell+1}) = \frac{t_{11}u_{\ell+1} + t_{12}}{t_{21}u_{\ell+1} + t_{22}}. \tag{1.275}$$

Sufficiency follows from the following observation. The product of $\Theta(s)$ and $G(s)$ in (1.255) is stable by definition. Hence multiplying this by $s - \alpha_i$ and letting $s \to \alpha_i$ eliminates all but one term $(1, -\beta_i)\Theta(\alpha_i)$, which is 0, implying that

$$t_{11}(\alpha_i) = \beta_i t_{21}(\alpha_i), \quad t_{12}(\alpha_i) = \beta_i t_{22}(\alpha_i). \tag{1.276}$$

By (1.262), this gives the interpolation condition (1.248) save for the last two (1.274). This is also satisfied since $u_{\ell+1}$ satisfies Condition (1.283), which is equivalent to (1.274). Then by Lemma 1.6 we may conclude that $u_1 \in \mathrm{BH}^\infty$ and whence the sufficiency follows.

Now we prove necessity. It suffices to show that the J-lossless $\Theta_k(s)$ cancels the unstable pole s_k. For this, a modification of [Kimura (1997), (5.27), p. 117] works, which reads in our situation

$$\frac{1}{s-\alpha_i}(1 \ -\beta_i)\Theta_k(s) = \left(\frac{\nu_i}{s+\bar{\alpha}_k} \ \ 0\right) + \frac{\mu_i}{s-\alpha_i}(1 \ -\beta_i'), \tag{1.277}$$

where

$$\nu_i = \frac{1-\beta_i\bar{\beta}_k}{\sqrt{1-|\beta_k|^2}}\frac{\alpha_i+\bar{\alpha}_k}{\alpha_k+\bar{\alpha}_k}, \quad \mu_i = \frac{1-\beta_i\bar{\beta}_k}{\sqrt{1-|\beta_k|^2}}\frac{\alpha_i-\alpha_k}{\alpha_i+\bar{\alpha}_k}. \tag{1.278}$$

Hence in particular,

$$\mu_k = 0 \tag{1.279}$$

i.e., Θ_k cancels the unstable pole at α_k. Hence Θ in (1.273) cancels all the unstable poles of $G = (A, B)$.

Moreover, from the very definition (1.254), it follows that

$$\Theta_k^*(j\omega)J\Theta_k(j\omega) = J, \tag{1.280}$$

whence

$$\Theta^*(j\omega)J\Theta(j\omega) = J, \tag{1.281}$$

i.e. Θ is J-lossless.

(1.281) implies as in [Kimura (1984)]

$$|f_1(j\omega)| = 1, \; |f_2(j\omega)| = |f_3(j\omega)| < 1, \; f_2(j\omega) = \bar{f}_3(j\omega). \tag{1.282}$$

By (1.271), (1.274) and (1.282), as in [Kimura (1984)] we conclude (1.266).

Hence, by Lemma 1.5, we may choose a suitable $u_{\ell+1}(s) \in \mathrm{BH}^\infty$ such that

$$u_{\ell+1}(0) = \delta_0, \quad u_{\ell+1}(\infty) = \delta_\infty \tag{1.283}$$

as in (1.265), then $u_1(s)$ satisfies (1.274) since (1.262) and (1.263) are equivalent.

□

Remark 1.14. Since A in Theorem 1.25 is anti-stable, Lemma 1.4 implies that the solution X of the Riccati equation (1.256) has the inverse $P = X^{-1}$ satsifying the Lyapnov type equation

$$AP + PA^* = BJB^*. \tag{1.284}$$

It can be seen readily that the solution of (1.284) is identical to the Pick matrix, so that Theorem 1.25 is another expression of Theorem 1.24 plus the additional interpolation condition (1.274).

1.12.3 *Proof of Theorem 1.23*

First we quote the standard result from [Doyle and Stein (1981)]. Σ indicates the unity feedback system Σ in Definition 1.8.

The synthesis problem of a controller C for the system Σ in Definition 1.8 refers to the sensitivity reduction problem, which asks for the estimation of the sensitivity function $S = S(s)$ multiplied by an appropriate frequency weighting function $W = W(s)$. First we show that

$$S = (I + PC)^{-1}, \tag{1.285}$$

is a transfer function from r to e.

Denoting the Laplace transforms by the corresponding capital letters, we have

$$Y = P(R - E),\ E = CY,$$

whence $Y = PR - PCY$. Solving in Y, we deduce that $Y = (I + PC)^{-1}PR$.

PC being the open loop transfer function, we have SR being the tracking error for the input R. Hence (1.285) holds true.

[Zames and Francis (1992)] introduced the function

$$q(s) = c(s)(1 + p_0(s)c(s))^{-1} \tag{1.286}$$

with nominal function $p_0(s)$ introduced in Definition 1.7, which can be solved for the stabilizer $c(s)$:

$$c(s) = q(s)(1 - p_0(s)q(s))^{-1}. \tag{1.287}$$

Lemma 1.7. *The controller $c(s)$ is a robust stabilizer for the class $C(p_0, r)$ if and only if Σ is stable for $p = p_0$ and*

$$|r(j\omega)q(j\omega)| < 1 \tag{1.288}$$

for all $\omega \in \mathbb{R}$.

Lemma 1.8. *The unity feedback system Σ is stable for $p = p_0$ if and only if* (i) *$q(s)$ is stable.*
(ii) *$1 - p_0(s)q(s)$ has the zeros at the unstable poles of $p_0(s)$ up to multiplicity. This implies that all the unstable poles of $p_0(s)$ are canceled out by the zeros of $q(s)$.*

We are now in a position to prove Theorem 1.23. Suppose $\alpha_1, \ldots, \alpha_\ell$ are all the unstable poles of $p_0(s)$. Let $B(s)$ be the product of all $B_i(s)$'s in (1.251)—the **Blaschke product** (for which see e.g. [Miller (1970)] and (1.153) above)

$$B(s) = \prod_{i=1}^{\ell} B_i(s), \tag{1.289}$$

which is a BH$^\infty$-function satisfying $|B(j\omega)| = 1$ and cancels all the unstable poles of $p_0(s)$. We modify $p_0(s)$ by eliminating all the poles by multiplying the BH$^\infty$-function ([Kimura (1984), (35]):

$$\tilde{p}_0(s) := sB(s)p_0(s) \tag{1.290}$$

is a stable function. Coupled with this, we introduce the function $\tilde{q}(s)$ so that we keep the relation

$$\tilde{p}_0(s)\tilde{q}(s) = p_0(s)q(s), \tag{1.291}$$

i.e.

$$\tilde{q}(s) := \frac{q(s)}{sB(s)} = \frac{p_0(s)q(s)}{\tilde{p}_0(s)}. \tag{1.292}$$

If $c(s)$ is a stabilizer of Σ, then by Lemma 1.8, (ii), $\tilde{q}(s)$ is also stable and furthermore, we have

$$\tilde{p}_0(\alpha_i)\tilde{q}(\alpha_i) = 1,\ 1 \leq i \leq \ell. \tag{1.293}$$

By the definition (1.286), we also have

$$\tilde{p}_0(0)\tilde{q}(0) = 1. \tag{1.294}$$

Defining

$$u(s) := r_m(s)\tilde{q}(s) = \frac{r(s)q(s)}{B(s)}, \tag{1.295}$$

we see that $|u(s)| = |r(s)q(s)|$ and so from (1.288) and the maximum modulus principle (Corollary 2.8) that $u(s)$ is a BH^∞ function. Also from (1.292), (1.293) and (1.294) we have

$$\beta_i := u(\alpha_i) = r'_m(\alpha_i)\tilde{q}(\alpha_i) = \frac{r'_m(\alpha_i)}{\tilde{p}_0(\alpha_i)},\ 1 \leq i \leq \ell \tag{1.296}$$

with the additional condition

$$\beta_0 := u(0) = \frac{r'_m(0)}{\tilde{p}_0(0)}. \tag{1.297}$$

Thus we are led to the GNP problem whose solution is given by Theorem 1.25.

1.12.4 *Another statement*

Theorem 1.26. *Suppose the notation and conditions in Theorem 1.23 all hold true. I.e.*

(A1) *All the unstable poles of $p_0(s)$ are simple and lie in $\mathcal{RHP}$ and $p_0(s)$ has a simple pole at $s = 0$.*

(A2) *The relative degree of the denominator of $r(s)$ is at most* 1.

Then the class of plants are robustly stabilizable if and only if the pair of matrices

$$A = \begin{pmatrix} \alpha_1 & 0 & \cdots & 0 \\ 0 & \alpha_2 & \cdots & 0 \\ & & \cdots & \\ 0 & 0 & \cdots & \alpha_\ell \end{pmatrix}, \quad B = \begin{pmatrix} 1 & -\beta_1 \\ 1 & -\beta_2 \\ & \cdots \\ 1 & -\beta_\ell \end{pmatrix} \tag{1.298}$$

has a J-lossless stabilizing conjugator Θ *for the signature matrix* (1.270) *and*

$$|\beta_0| = |r_m'(0)/\tilde{p}_0(0)| < 1. \tag{1.299}$$

Example 1.38. Numerical data is given in [Ban *et al.* (2009)].

$$a = 10.8, b = 228, \tag{1.300}$$
$$a = 8.64, b = 274,$$

but we shall not dwell on this side. Leaving the design of the robust controller ([MacFarlane and Glover (1996)]) for another occasion.

1.13 Paley-Wiener theorem

The material in this section will be further elaborated in §2.14.

In this section we assemble basics of the Paley-Wiener theorem according to Rudin [Rudin (1986)]. There are two classes of functions that are analyzed and what is relevant to us is the one similar to (2.227). There are some books and papers related to this theorem (cf. e.g. [Takada (1985), pp. 38-40]).

$$f(z) = \frac{1}{\sqrt{2\pi}} \int_{-A}^{A} F(\omega) e^{i\omega z} \, \mathrm{d}\omega, \tag{1.301}$$

where $A > 0$ and $F \in L^2(-A, A)$. One can show that f is an integral function (Definition 1.3). It satisfies the growth condition

$$|f(z)| \le \frac{1}{\sqrt{2\pi}} \int_{-A}^{A} |F(\omega)| e^{-\omega y} \, \mathrm{d}\omega \le \frac{1}{\sqrt{2\pi}} e^{A|y|} \int_{-A}^{A} |F(\omega)| \, \mathrm{d}\omega, \tag{1.302}$$

where we write $z = x + iy$. Hence denoting the last integral by C, we obtain

$$|f(z)| \le C e^{A|z|}. \tag{1.303}$$

Definition 1.9. An integral function f that satisfies condition (1.303) is said to be **of exponential type** (or of order 1 a là Hadamard).

Thus we have

Proposition 1.4. *Every f of the form* (1.301) *is an integral function which satisfies* (1.303) *and whose restriction is in L^2 (by the Plancherel theorem).*

The remarkable theorem of Paley-Wiener (Theorem 2.34) asserts that the converse also holds, which we state in a more detailed form.

Theorem 1.27. (Paley-Wiener theorem) *Suppose A and C are positive constants, that f is an* **integral function of exponential type,** *i.e. satisfying* (1.303) *for all values of z, and that*

$$\int_{-\infty}^{\infty} |f(z)|^2 < \infty. \tag{1.304}$$

Then there exists an $F \in L^2(-\infty, \infty)$ such that

$$f(z) = \frac{1}{\sqrt{2\pi}} \int_{-A}^{A} F(\omega) e^{i\omega z}\, d\omega \tag{1.305}$$

for **all** *values of z.*

1.14 Bernstein polynomials

1.14.1 *Time-limited polynomial extrapolation*

If f is an analytic function, then the partial sums of its Taylor series in the domain D of analyticity are the sequence of polynomials that converge to f, uniformly on any compact subset of D. The problem of extrapolation relies on the integrality of the function f, which we so suppose in what follows. Then the above statement becomes valid by changing D into $\mathbb{C}$. The point is that we may find other polynomial approximations to f as well. We adopt the argument of [Giardina (1984)] using the Bernstein polynomials. For a general theory of Bernstein polynomials, we refer to [Lorentz (2012)].

Let f be a band-limited signal observed on the interval $[0, 1]$. Then as with the sampling theorem, we may use the $m + 1$ equally spaced samplings of $[0, 1]$ to construct a sequence of polynomials of degree $\le m$ in the following

Theorem 1.28. *Let f be an integral function and let $g_m(t)$ denote the* **time-limited signals** *defined by*

$$g_m(t) = \sum_{k=0}^{m} f\left(\frac{k}{m}\right) B_k^m(t) \chi_{[-m,m]}(t), \tag{1.306}$$

where $\chi_{[-m,m]} = \chi_{[-m,m]}(t)$ *indicates the characteristic function of the interval* $[-m, m]$, *which entails the time-limitedness of the signal* $g_m(t)$. *Then* $g_m(t)$ *converges to* f *on* $\mathbb{R}$ *and the convergence is uniform on any compact set.*

Furthermore, the functions

$$G_{m,T}(\omega) = 2\sum_{k=0}^{m} f\left(\frac{k}{m}\right) \sum_{j=0}^{m} \frac{(-1)^j}{(-i)^{k+j}} \frac{\mathrm{d}^{k+j} \frac{\sin \omega T}{\omega}}{\mathrm{d}\omega^{k+j}} \binom{m-k}{j} \tag{1.307}$$

can be used as an approximation to the Fourier transform of the band-limited signal $f = f(t)$. *If* f *is in* L^1, *then* $G_{m,T}(\omega)$ *converges uniformly to* $\hat{f}$.

Here $B_k^m(t)$ are the **Bernstein basis functions**

$$B_k^m(x) = \binom{m}{k} x^k (1-x)^{m-k} \tag{1.308}$$

the addends for the **Bernstein polynomial**

$$\mathcal{P}_n(x) = \sum_{k=0}^{n} c_k^n B_k^n(x). \tag{1.309}$$

The special case of $c_k^n = f\left(\frac{k}{n}\right)$ is usually used which we state

$$\mathcal{B}_n(f,t) = \sum_{k=0}^{n} f\left(\frac{k}{n}\right) B_k^n(t) = \sum_{k=0}^{n} f\left(\frac{k}{n}\right) \binom{n}{k} t^k (1-t)^{n-k}. \tag{1.310}$$

For a proof of Theorem 1.28, the following theorem of Kantrovich is used.

Theorem 1.29. *If* f *is an integral function, then*

$$\lim_{m\to\infty} \mathcal{B}_m(f(t), t) = f(t) \tag{1.311}$$

for all real values of t *and the convergence is uniform on any compact interval* $[-R, R]$ *for ant* $R > 0$.

1.15 Some far-reaching principles in mathematics

(i) The **formalism-preserving principle** or the **embedding principle** refers to the principle to interpret a given system S by *embedding* it in a wider one $\mathcal{S}$ while preserving all the formalism holding in S, to the effect that older formalism S still holds true in the new framework as formalism in $\mathcal{S}$. This sometimes works as guiding principle for constructing new systems. A good example is the tower of number systems in (2.32). The $x \to z$-principle (or line $\to$ plane-principle) stated after Theorem 2.5 is another example. In this an endeavor is made to extend the domain of definition of an analytic function into a wider range while keeping analyticity. Consistency theorem (Theorem 2.2) says that changing the variable from x to z is the only way to extend a real analytic function into a complex analytic function. Consistency theorem also belongs to the following uniqueness principle.

(ii) The **uniqueness principle** refers to the uniqueness of an ingredient satisfying the prescribed conditions is unique. In the case of inversion, it says that given an operator, the inverse operator, if it exists, is uniquely determined up to a certain extra ingredient. E.g. Exercise 2.9 shows the uniqueness of the inverse element of any element in a group. (2.62) states the (almost) inversion of differentiation and integration.

The principle entails that in most of the expansions, the coefficients are uniquely determined. The expansions one learns include Taylor, Laurent, Fourier expansion. The method of undetermined coefficients is effectively used in the first two expansions and in subsequent sections on partial fraction expansions. Indeed, we may say that one of the main purposes as well as a unique feature of our book is this aspect of recourse to partial fraction expansions. Uniqueness of Fourier coefficients follow from the orthogonality.

In solving the DE, the inverse Laplace transform in Corollary 1.7 is effectively used in § 1.11.2. It's a variation of the Fourier inversion (2.214) which makes it possible to move between *time domain* and *frequency domain*. This continues to the following.

(iii) The **local-global principle** to the effect that collecting local data gives rise to the totality $\sum \mathrm{d}A = A$. In the definition of various integrals, this appears in the form with $\mathrm{d}A = f(x)\,\mathrm{d}x$:

$$\int \mathrm{d}A = \int_a^b f(x)\,\mathrm{d}x = A.$$

Here again, the molecular-biological philosophy is to be taken into account with care that "genotype determines phenotype". In rigid systems the local-global principle holds true but in a more fluctuating system such as living organisms this is to be applied up to some allowance. The same applies here to the effect that the analogue input into digital apparatus, being processed by a certain digital device and released as analogue output, often described as $A/D \to D/A$ in engineering books should be $A/D \to D/A'$

with some ambiguities. Since we perceive the output analogue data by our own perceptive organs, they are deceived to some extent or can recognize it as meaningful analogue data. The partial fraction expansion in (ii) is also a sort of local-global principle for which one needs the Liouville theorem as to reach coincidence.

(iv) The **reduction principle** means the classification of objects modulo a certain standard or identifying two different objects which are different up to some ignorable entity. In Example 2.1 reduction modulo 5 appears. In the (Banach) space $L^p(D)$ of all p-th power integrable functions, each element is not a function itself but an equivalence class modulo a measure 0 set.

[illegible] ambiguities. Since we perceive the output and [illegible] it by our own perceptual organs, they are deceived to the extent of [illegible] recognize it as meaningful [illegible]. The normal function expressed in (h) is also a [illegible] principle which [illegible] the [illegible] to [illegible].

(c) The reduction principle means the classification of objects in a [illegible] set are identical or identifying two different objects which [illegible] differ up to some isotopic curve. In Example 2.1 [illegible] modulo [illegible]. In the (Banach) space L^2 [illegible] of [illegible] functions, each element is not a function itself but an equivalence class modulo measure 0 set.

Chapter 2

Applicable Real and Complex Functions

This chapter serves for founding solid basics for real and complex analysis abridging a gap between them, incorporating many applications of real analysis. An advanced reader can skip those parts which are familiar to him.

2.1 Preliminaries

We denote the set of all natural numbers by $\mathbb{N} = \{1, 2, 3, \ldots\}$.

"Mathematical induction" refers to the principle of proving a proposition with regard to all natural numbers $\geq n_0 \in \mathbb{N}$ by proving first that

(i) it is true for $n_0 \in \mathbb{N}$,

and then that

(ii) if we assume its validity for $(n_0 \leq) \, n \in \mathbb{N}$, then it is valid also for $n+1$.

Instead of (ii) we may adopt either

(ii') if we assume the proposition is valid for all natural numbers $\leq n$, then it is also valid for $n+1$,

or changing the variable,

(ii") if we assume it is valid for $n-1$, then it is also valid for n.

Exercise 2.1. For x_1, x_2, $\ldots$, $x_n \in \mathbb{R}$ (real numbers, cf. §2.2.1), prove the **triangular inequality** by induction: $\left|\sum_{k=1}^{n} x_k\right| \leq \sum_{k=1}^{n} |x_k|$. Also, locate the case where the equality holds true. Cf. Exercise 2.18.

Solution It suffices to prove the case of $n = 2$. The first proof is peculiar to 1-dimensional and one cannot step further. If $x_1 x_2 \geq 0$, then the equality holds. If $x_1x_2 < 0$, then we may suppose that $|x_1| > |x_2| > 0$, in which case $|x_1 + x_2| = |x_1| - |x_2| < |x_1| < |x_1| + |x_2|$.

The second proof is feasible for generalization:

$$\begin{aligned} |x_1 + x_2|^2 = (x_1 + x_2)^2 = |x_1|^2 + |x_2|^2 + 2x_1x_2 &\le |x_1|^2 + |x_2|^2 + 2|x_1x_2| \\ = |x_1|^2 + |x_2|^2 + 2|x_1||x_2| &= (|x_1| + |x_2|)^2. \end{aligned}$$

Exercise 2.2. Prove the following binomial theorem by mathematical induction.

Binomial theorem. Provided that the product of a and b is commutative, we have

$$(a+b)^n = \sum_{k=0}^{n} \binom{n}{k} a^{n-k} b^k, \tag{2.1}$$

where $\binom{n}{k} = {}_nC_k = \frac{n!}{k!(n-k)!} = \frac{n(n-1)\cdots(n-k+1)}{k!}$ is called the **binomial coefficient** expressing the number of combinations of choosing r objects from n objects. $n! = 1 \cdot 2 \cdot \cdots \cdot n$ is called the **factorial** of n. [The identity $\binom{z}{k} = \frac{z(z-1)\cdots(z-k+1)}{k!}$ holds for all $z \in \mathbb{C}$ when viewed as the definition of a polynomial of degree k for $0 \le k \in \mathbb{Z}$.]

Exercise 2.3. For $n \in \mathbb{N}$ prove the following by induction. Prove (i) and (ii) by the binomial theorem.

(i) $2^n > n$,

(ii) if $h \ge -1$, then $(1+h)^n \ge 1 + nh$ (Bernoulli's inequality).

Exercise 2.4. Prove that if $X \subset \mathbb{R}$ and $\#X < \infty$ (finitely many elements), then X has both the minimal element $\min X$ and the maximal element $\max X$.

Exercise 2.5. If $\mathbb{N} \supset X \neq \varnothing$, then there is the least element of X.

Solution. Suppose X has no minimal element. Then $1 \notin X$ since 1 is the least natural number. Suppose then that $1, \ldots, k-1 \notin X$. Then $k \notin X$ since if $k \in X$, then k would be the minimal element of X. Hence by induction, we conclude that $X = \varnothing$, a contradiction.

Exercise 2.5 is often applied in proving the existence of a generator.

Theorem 2.1. *The* **sum formula for a geometric progression**

$$(1-r)\sum_{k=0}^{n-1} r^k = 1 - r^n, \quad \sum_{k=0}^{n-1} r^k = \frac{1-r^n}{1-r} \quad (r \neq 1) \tag{2.2}$$

and the factorization formula

$$(a-b)\sum_{k=0}^{n-1} a^{n-k-1}b^k = a^n - b^n \tag{2.3}$$

are equivalent.

Proof follows under the substitution $r = \frac{b}{a}$.

Exercise 2.6. Suppose $r \neq 1$ and $r^q = 1$ for $2 \leq q \in \mathbb{N}$. Then prove that

$$\sum_{n=1}^{q} nr^n = \frac{qr}{r-1} \tag{2.4}$$

and that

$$\sum_{n=1}^{q-1} n(r^n - r^{-n}) = q\frac{r^{1/2}+r^{-1/2}}{r^{1/2}-r^{-1/2}}. \tag{2.5}$$

Exercise 2.7. For $w \neq z$, $w \neq z_0$, we have

$$\frac{1}{w-z} = \sum_{k=0}^{n-1} \frac{(z-z_0)^k}{(w-z_0)^{k+1}} + \frac{1}{w-z}\left(\frac{z-z_0}{w-z_0}\right)^n. \tag{2.6}$$

Solution We write $\frac{1}{w-z} = \frac{1}{w-z_0+z_0-z} = \frac{1}{w-z_0}\frac{1}{1-\frac{z-z_0}{w-z_0}}$ and use

$$\frac{1}{1-r} = \frac{1-r^n}{1-r} + \frac{r^n}{1-r} = \sum_{k=0}^{n-1} r^k + \frac{r^n}{1-r},$$

where for the first term on the right we used (2.2), with $r = \frac{z-z_0}{w-z_0}$ to deduce that

$$\frac{1}{w-z} = \frac{1}{w-z_0}\left(\sum_{k=0}^{n-1}\left(\frac{z-z_0}{w-z_0}\right)^k + \frac{\left(\frac{z-z_0}{w-z_0}\right)^n}{1-\frac{z-z_0}{w-z_0}}\right).$$

Further, if $\left|\frac{z-z_0}{w-w_0}\right| = |r| < 1$, then $\frac{1}{w-z} = \sum\limits_{k=0}^{\infty} \frac{(z-z_0)^k}{(w-z_0)^{k+1}}$, the series being absolutely and uniformly convergent in w and z, which will be used in the proof of the Laurent expansion.

2.2 Algebra of complex numbers

2.2.1 *Algebraic preliminaries and embeddings*

Definition 2.1. A (non-empty) set K is called a **field** if there are defined four (binary) operations in arithmetic-addition, subtraction, multiplication and division (save for one by 0) and these can always be conducted within K.

More precisely, we suppose the following. There are two binary operations, called addition and multiplication, $+$ and $\cdot$ defined in K satisfying
(0) if $a, b \in K$, then $a + b \in K$ and $a \cdot b \in K$ (for simplicity we write ab for $a \cdot b$)
(i) (associative law). For $a, b, c \in K$,

$$(a+b)+c = a+(b+c), \quad (ab)c = a(bc)$$

(ii) (existence of identity (unit) element). There are two special elements 0 and 1 such that for any $a \in K$

$$a+0 = 0+a = a, \quad a1 = 1a = a.$$

(iii) (existence of inverse elements). For each a there is the additive inverse $-a$ such that

$$a + (-a) = -a + a = 0;$$

for each $a \neq 0$, and there is the multiplicative inverse a^{-1} such that

$$aa^{-1} = a^{-1}a = 1. \tag{2.7}$$

(iv) (distributive law). For $a, b, c \in K$,

$$a(b+c) = ab + ac, \quad (a+b)c = ac + bc.$$

(v) (commutative law). For $a, b \in K$,

$$a + b = b + a, \quad ab = ba.$$

Remark 2.1. We define subtraction as the inverse operation of addition:

$$a - b = a + (-b),$$

and division as one of multiplication:

$$a \div b = ab^{-1}$$

the principle of Ying and Yang.

There are essentially two operations, addition and multiplication, and (iv) determines the relation between $+$ and $\cdot$. (i) is also essential in the sense that they claim the result of two ways of operations, a and b first and then c, and b and c first and then a, is the same. Without this rule, there would occur an excessive complication.

Exercise 2.8. Prove that there is only one identity element.

Exercise 2.9. (Uniqueness of the inverse) Prove that for each a, there is a unique inverse element.

Solution. If there is another element a' satisfying the same equality as (2.7):

$$aa' = a'a = e.$$

Then $a' = a'e = a'aa^{-1} = ea^{-1} = a^{-1}$.

Exercise 2.10. Prove that if a and b have inverses, then

$$(ab)^{-1} = b^{-1}a^{-1}. \tag{2.8}$$

Solution. Since $b^{-1}a^{-1}ab = abb^{-1}a^{-1} = e$, we may appeal to Exercise 2.9 to conclude Exercise 2.8.

Exercise 2.11. Use the above rule to deduce that for all $a \in K$, $a0 = 0a = 0$.

Solution. Putting $b = c = 0$ in the distributive law in Definition 2.1, we get $a0 = a0 + a0$. Adding $-a0$ to both sides proves the assertion.

Exercise 2.12. Explain the reason why you cannot divide by 0.

Solution. Suppose division by 0 is possible in K in the sense of Remark 2.1, i.e. suppose that the inverse element 0^{-1} exists in K. Then $00^{-1} = 0^{-1}0 = 1$. However, by Exercise 2.11, $0^{-1}0 = 0$, whence we conclude that $1 = 0$. Hence for any $a \in K$, we have $a = a1 = a0 = 0$, i.e. $K = \{0\}$, a singleton, which is usually ruled out from consideration. Recently there is a new theory being developed of absolute mathematics over this one-element field.

In the set of natural numbers, only addition and multiplication are always possible. Therefore we extend the range of $\mathbb{N}$ to construct the set $\mathbb{Z}$ of (rational) integers:

$$\mathbb{Z} = \{0, \pm 1, \pm 2, \ldots\}$$

in which addition and subtraction (multiplication) are possible and $\mathbb{Z}$ forms a ring. A ring is nearly a field with division not always possible. The ring $\mathbb{Z}$ of integers is of particular type being a Euclidean domain, i.e. a domain in which the Euclidean division is possible.

Definition 2.2. (Euclidean division) Given $m \in \mathbb{Z}$ and $n \in \mathbb{N}$, there exist $q, r \in \mathbb{Z}$ such that

$$m = nq + r, \quad 0 \le r < n, \tag{2.9}$$

where q is called the quotient and r the (least positive) residue modulo n.

Exercise 2.13. (i) Let $[x]$ denote the greatest integer not exceeding $x \in \mathbb{R}$, called the **integer part** or the Gaussian symbol, of x. Then prove that

$$x - 1 < [x] \le x. \tag{2.10}$$

(ii) Prove that the quotient and residue in (2.9) are given respectively by $q = \left[\frac{m}{n}\right]$ and $r = m - \left[\frac{m}{n}\right] n$.

Solution. (i) Since $[x]$ does not exceed x, we have the second inequality. Suppose $x - 1 \ge [x]$. Then $[x] + 1 \le x$. Since $[x] + 1$ is an integer bigger than $[x]$, this contradicts the maximality of $[x]$.
(ii) Putting $q = \left[\frac{m}{n}\right]$, then by (i), $\frac{m}{n} - 1 < q \le \frac{m}{n}$, or $m - n < nq \le m$, so that $r = m - nq$ satisfies (2.9).

We extend $\mathbb{Z}$ to obtain the field $\mathbb{Q}$ of rational numbers:

$$\mathbb{Q} = \left\{ \frac{m}{n} \middle| m, n \in \mathbb{Z}, n \ne 0 \right\},$$

in which all four operations are possible and is sufficient for everyday arithmetic.

However, when you consider irrational numbers, the range of $\mathbb{Q}$ is not wide enough e.g. you cannot think of the number whose square is 2, which was known to ancient Babylonians.

Exercise 2.14. Prove that $\sqrt{2} \notin \mathbb{Q}$.

Finding a root ($\sqrt{2}$) of the polynomial $X^2 - 2 \in \mathbb{Q}[X]$ (the polynomial ring over $\mathbb{Q}$) is equivocal to the decomposition into irreducible polynomials (in this case, into the product of linear factors), and it is natural to try to find the smallest range (the field of coefficients) over which $X^2 - 2$

decomposes into linear factors. It is easy to find such a field. It is the field obtained by adjoining $\sqrt{2}$ to $\mathbb{Q}$

$$\mathbb{Q}(\sqrt{2}) = \{a + b\sqrt{2} \mid a, b \in \mathbb{Q}\}.$$

Exercise 2.15. Prove that $\mathbb{Q}(\sqrt{2})$ forms a field (which is called a **real quadratic field**).

In the **polynomial ring** $\mathbb{Q}(\sqrt{2})[X]$ (i.e. in the ring of all polynomials with coefficients in $\mathbb{Q}(\sqrt{2})$), we have the decomposition

$$X^2 - 2 = (X - \sqrt{2})(X + \sqrt{2})$$

and from the point of view of the prime decomposition, $\mathbb{Q}(\sqrt{2})$ is enough for this particular polynomial $X^2 - 2$.

Similarly, for the roots $\alpha_1, \ldots, \alpha_n$ of a polynomial $f(X) \in \mathbb{Q}[X]$ of degree n, we may construct the **splitting field** $\mathbb{Q}(\alpha_1, \ldots, \alpha_n)$ and the prime decomposition is possible in $\mathbb{Q}(\alpha_1, \ldots, \alpha_n)[\mathrm{X}]$. This type of numbers is called **algebraic numbers**. There are, however, much more non-algebraic (called **transcendental**) numbers, such as e, π, $\log \pi$ etc. What should we do with these numbers out of control of algebraic operations?

In developing analysis, the fundamental concept is the "limit", and the following case brings about a more serious problem. Consider the sequence a_n of rational numbers generated by the rule

$$a_n = \frac{1}{2}\left(a_{n-1} + \frac{2}{a_{n-1}}\right), \quad n \geq 2$$

starting from any rational value $a_1 = 1$, say. Then $a_2 = \frac{3}{2} = 1.5$, $a_3 = \frac{17}{12} = 1.417$.

Exercise 2.16. Prove that $\lim_{n\to\infty} a_n = \sqrt{2}$.

The situation being that all terms are rational but the limit is not rational, is compared to a traveler tracing the sound step of rationals and encountering a pit at the furthest end of the path, a non-welcoming outcome.

This is caused by the fact that the rationals do not exhaust the whole line but there are many holes which are not labeled by rationals.

To overcome this difficulty, we are to construct a number system (that of real numbers) $\mathbb{R}$ that corresponds one to one to all points on the line (i.e. the real numbers *label* all points on the line). The construction of such a number system is not easy at all and we must take the existence of real numbers for granted in this book.

In the 19th century many distinguished mathematicians devoted a considerable amount of time to construct a real number system; Dedekind, Weierstrass, Cauchy, Cantor *et al.* Once constructed, the system must have an axiom that describes the continuity of a line in term of the properties of the real numbers in the system, which is called the "continuity of real numbers".

The method of Cauchy sequences due to Cauchy for constructing real number system is one of the most efficient one–all sequences modulo null sequences–since it gives rise to a **complete set**, i.e. all Cauchy sequences are convergent.

$\mathbb{R}$ being complete, all Cauchy sequences are convergent, and limit process is legitimate.

2.2.2 *Complex number system*

There was another objective of extending the range of $\mathbb{Q}$, like constructing $\mathbb{Q}(\sqrt{2})$, i.e. of securing the decomposition into prime factors. $\mathbb{R}$ is not enough for this purpose, for even the simplest quadratic polynomial $aX^2 + bX + c$, $(a \neq 0)$ with discriminant $D = b^2 - 4ac < 0$ cannot be decomposed in $\mathbb{R}[X]$.

In order to assure the prime decomposition we need to introduce an imaginary "number" whose square is -1. This cannot be a real number because a square of a real number is ≥ 0. We need to construct a new number system which contains $\mathbb{R}$ and in which prime decomposition is always possible. The remarkable construction of Gauss is given as the first construction in Example 1.3 in § 1.2 (we review the construction in Exercise 2.17) by which we think of a two-dimensional real vector $\boldsymbol{z} = \begin{pmatrix} x \\ y \end{pmatrix}$ as a complex number $z = x + iy$ (recall Figure 1.3):

$$\boldsymbol{z} = \begin{pmatrix} x \\ y \end{pmatrix} \in \mathbb{R}^2 \longleftrightarrow z = x + iy \in \mathbb{C}. \tag{1.9}$$

Exercise 2.17. We introduce the special operation $*$ between two vectors $\boldsymbol{z}_i = \begin{pmatrix} x_i \\ y_i \end{pmatrix}$, $i = 1, 2$ in $\mathbb{R}^2$ defined by

$$\boldsymbol{z}_1 * \boldsymbol{z}_2 = \begin{pmatrix} x_1 \\ y_1 \end{pmatrix} * \begin{pmatrix} x_2 \\ y_2 \end{pmatrix} = \begin{pmatrix} x_1 x_2 - y_1 y_2 \\ x_1 y_2 + x_2 y_1 \end{pmatrix}. \tag{2.11}$$

(i) Show that with this $*$ as multiplication and the ordinary addition of vectors $\boldsymbol{z}_1 + \boldsymbol{z}_2 = \begin{pmatrix} x_1 + x_2 \\ y_1 + y_2 \end{pmatrix}$ as addition, $\mathbb{R}^2$ forms a field in the sense of Definition 2.1.

(ii) Show that a vector $\boldsymbol{z} = \begin{pmatrix} x \\ y \end{pmatrix}$ may be thought of as a "complex number" (or a symbol) $x + iy$, where i (the **imaginary unit**) is the special vector $\begin{pmatrix} 0 \\ 1 \end{pmatrix}$ whose square with respect to $*$ is $\begin{pmatrix} -1 \\ 0 \end{pmatrix}$. Check that if we multiply out $z_1 z_2 = (x_1 + iy_1)(x_2 + iy_2)$ as we do with numbers and replace i^2 by -1, then we get the "number" corresponding to the vector $\boldsymbol{z}_1 * \boldsymbol{z}_2$.

(iii) Show that the real number system $\mathbb{R} = \{x\}$ may be identified with the vectors $\left\{ \begin{pmatrix} x \\ 0 \end{pmatrix} \right\}$, which is known as an embedding.

(iv) Show that the complex number system $\mathbb{C} = \{z = x + iy | x, y \in \mathbb{R}\}$ forms a field.

By Exercise 2.17, we may work with $z = x + iy$ just as we do with numbers with one convention that the whenever we encounter i^2, we are to replace it by -1:

$$i^2 = -1. \tag{2.12}$$

Under the identification (1.9) we call the x-coordinate the **real part** of z (denoted $\operatorname{Re} z$) and the y-coordinate the **imaginary part** ($\operatorname{Im} z$). Equality of two complex numbers $z_1 = x_1 + iy_1$ and $z_2 = x_2 + iy_2$ is the same as that of real and imaginary parts, being Cartesian coordinates:

$$\begin{aligned} x_1 + iy_1 = x_2 + iy_2 &\iff x_1 = x_2, \; y_1 = y_2, \\ z_1 = z_2 &\iff \operatorname{Re} z_1 = \operatorname{Re} z_2, \; \operatorname{Im} z_1 = \operatorname{Im} z_2. \end{aligned} \tag{2.13}$$

Geometrically, $\mathbb{C}$ is the plane $\mathbb{R}^2$ with the x-axis (called the **real axis**) the same as $\mathbb{R}$ and the y-axis (**imaginary axis**) with scales iy instead of $y \in \mathbb{R}$, i.e. we think of each complex number $z = x + yi$ as the sum of two vectors $x = x \begin{pmatrix} 1 \\ 0 \end{pmatrix}$ and $yi = y \begin{pmatrix} 0 \\ 1 \end{pmatrix}$. Viewed in this way, the plane $\mathbb{R}^2$ is called the **complex plane** or the **Gauss plane** and is denoted by the same letter $\mathbb{C}$. We define the **absolute value** $|z|$ of z as

$$|z| = |x + iy| = \sqrt{x^2 + y^2}. \tag{2.14}$$

The mirror image of $z = x+iy$ relative to the real axis is called the **complex conjugate**, $\bar{z}$, of z:

$$\bar{z} = \overline{x + iy} = x - iy, \tag{2.15}$$

and the mapping $^{-}$ is called the complex conjugation. Since the distances of $\boldsymbol{z} = \begin{pmatrix} x \\ y \end{pmatrix}$ and $\bar{\boldsymbol{z}} = \begin{pmatrix} x \\ -y \end{pmatrix}$ from $\boldsymbol{o}$ is the same, we have $|\bar{z}| = |z|$.

We remark that $*$-operation in (2.11) is quite natural because

$$\bar{\boldsymbol{z}}_1 * \boldsymbol{z}_2 = \begin{pmatrix} x_1 \\ -y_1 \end{pmatrix} * \begin{pmatrix} x_2 \\ y_2 \end{pmatrix} = \begin{pmatrix} x_1x_2 + y_1y_2 \\ x_1y_2 - x_2y_1 \end{pmatrix}$$

and $x_1x_2 + y_1y_2 = \boldsymbol{z}_1 \cdot \boldsymbol{z}_2$ (inner product) and $x_1y_2 - y_1y_2 = \det(\boldsymbol{z}_1, \boldsymbol{z}_2)$ (determinant), cf. § 1.2.

We note that for $z = x + iy$

$$z\bar{z} = |z|^2 = |\bar{z}|^2 = |\boldsymbol{z}|^2 = x^2 + y^2, \tag{2.16}$$

where the fourth member is the length of the vector $\boldsymbol{z} \longleftrightarrow z$.

Exercise 2.18. Prove that

$$\mathbb{Q}(i) = \{a + bi \,|\, a, b \in \mathbb{Q}\}$$

forms a field (called the **Gaussian field**) and that $\mathbb{Q}(\rho) = \{a+b\rho \,|\, a, b \in \mathbb{Q}\}$ forms a field (called the **Eisenstein field**), where $\rho = \frac{-1+\sqrt{3}i}{2}$ is a primitive third root of 1.

Exercise 2.19. (i) Prove that the complex conjugation is a ring isomorphism from $\mathbb{C}$ into itself:

$$\overline{z_1 + z_2} = \overline{z_1} + \overline{z_2}, \quad \overline{z_1 z_2} = \overline{z_1}\,\overline{z_2} \tag{2.17}$$

(and consequently, $\overline{z_1 - z_2} = \overline{z_1} - \overline{z_2}$ and $\overline{\frac{z_1}{z_2}} = \frac{\overline{z_1}}{\overline{z_2}}$ for $z_2 \neq 0$).
(ii) Prove that $z + \overline{z} = 2\,\mathrm{Re}\, z$, $z - \overline{z} = 2i\,\mathrm{Im}\, z$ and that $z \in \mathbb{R} \iff z = \overline{z}$.
(iii) Prove that if $\alpha \notin \mathbb{R}$ is a root of the algebraic equation

$$a_0x^n + a_1x^{n-1} + \cdots + a_n = 0$$

with real coefficients, then $\overline{\alpha}$ is also a root of the same equation.

Exercise 2.20. (i) Prove the **multiplication formula** for the absolute value

$$|z_1z_2| = |z_1||z_2| \tag{2.18}$$

by the 2-dimensional **Lagrange formula**

$$(x_1^2 + y_1^2)(x_2^2 + y_2^2) = (x_1x_2 - y_1y_2)^2 + (x_1y_2 + x_2y_1)^2 \tag{2.19}$$

(which was used by Lagrange in the proof of his Two Squares Theorem to the effect that the product of two integers in the form of a sum of two squares is again a sum of two squares).

(ii) Prove (2.18) by (2.16) and (2.17).
(iii) Prove the **triangular inequalities**

$$||z_1| - |z_2|| \le |z_1 \pm z_2| \le |z_1| + |z_2|. \tag{2.20}$$

Cf. Exercise 2.1.
(iv) Generalize (2.20) to

$$\left|\sum_{k=1}^{n} z_k\right| \le \sum_{k=1}^{n} |z_k|, \quad z_k \in \mathbb{C},\ 1 \le k \le n.$$

Solution. We give two different proofs of (iii). The first proof rests on (2.16) and (2.19):

$$\begin{aligned}|z_1 + z_2|^2 &= (z_1 + z_2)\overline{(z_1 + z_2)} = z_1\overline{z_1} + 2\,\mathrm{Re}\, z_1\overline{z_2} + z_2\overline{z_2}\\ &= |z_1|^2 + |z_2|^2 + 2\,\mathrm{Re}\, z_1\overline{z_2} \le |z_1|^2 + |z_2|^2 + 2|z_1\overline{z_2}|\\ &\le |z_1|^2 + |z_2|^2 + 2|z_1||\overline{z_2}| = (|z_1| + |z_2|)^2\end{aligned}$$

whence the second inequality of (2.20) follows. The first inequality of (2.20) follows from it by subtraction.

The second proof follows from the Cauchy-Schwartz inequality:

$$|\boldsymbol{z}_1 \cdot \boldsymbol{z}_2| \le |\boldsymbol{z}_1||\boldsymbol{z}_2|. \tag{2.21}$$

Indeed,

$$\begin{aligned}|z_1 + z_2|^2 &= |\boldsymbol{z}_1 + \boldsymbol{z}_2|^2 = (\boldsymbol{z}_1 + \boldsymbol{z}_2)\cdot(\boldsymbol{z}_1 + \boldsymbol{z}_2) = |\boldsymbol{z}_1|^2 + |\boldsymbol{z}_2|^2 + 2\boldsymbol{z}_1 \cdot \boldsymbol{z}_2\\ &\le |\boldsymbol{z}_1|^2 + |\boldsymbol{z}_2|^2 + 2|\boldsymbol{z}_1||\boldsymbol{z}_2| = (|\boldsymbol{z}_1| + |\boldsymbol{z}_2|)^2 = (|z_1| + |z_2|)^2.\end{aligned}$$

Note that (2.21) is a consequence of (2.19), whence that it is the Lagrange identity (2.19) that underlies the triangular inequality and that (2.20) verifies that the distance function introduced in § 1.2 is indeed a distance function.

Exercise 2.21. This exercise is a matrix version of Example 1.4 and you will learn how simple it is to work with matrices than with Cartesian coordinates.

Prove that the set of all matrices of the form

$$Z = \begin{pmatrix} x & -y \\ y & x \end{pmatrix} \tag{2.22}$$

with $x, y \in \mathbb{R}$ forms a field with respect to ordinary addition and multiplication of matrices, and that it is isomorphic to the complex number field $\mathbb{C}$. Then check that (2.19) can be interpreted as the multiplicativity of the determinants: $|Z_1||Z_2| = |Z_1 Z_2|$.

In addition to the rectangular coordinates $\boldsymbol{z} = \begin{pmatrix} x \\ y \end{pmatrix}$ of the complex number $z = x + iy$, the **polar coordinates** $\begin{pmatrix} r \\ \theta \end{pmatrix}$ are of great use. If $\begin{pmatrix} r \\ \theta \end{pmatrix}$ is the polar coordinate of the point $\boldsymbol{z} = \begin{pmatrix} x \\ y \end{pmatrix} \neq \mathbf{o}$ then

$$\begin{cases} x = r\cos\theta, \\ y = r\sin\theta \end{cases} \tag{2.23}$$

and so

$$z = r(\cos\theta + i\sin\theta) = re^{i\theta}, \tag{2.24}$$

where the last equality is (1.29). This is called the **polar form** of $z \neq 0$ and $r = |z|$ is the absolute value of z in (2.14) and θ, denoted by $\arg z$(or amp z), is called the **argument** of z. At the origin $z = 0$, the absolute value is 0 and the argument is indefinite. Fixing a value θ_0 of $\arg z$, the argument is a multi-valued function expressed as

$$\arg z = \theta_0 + 2\pi n \quad (n \in \mathbb{Z}).$$

To handle the multi-valuedness of $\arg z$, we often restrict its range to one cycle $[0, 2\pi)$ or $(-\pi, \pi]$, in which case we may treat it as a single-valued function.

To find the polar form of $z = x + iy \neq 0$, we first calculate $r = |z| = \sqrt{x^2 + y^2}$. Then factor out r to get $z = r\left(\frac{x}{r} + i\frac{y}{r}\right)$.

Then find θ such that

$$\cos\theta = \frac{x}{r}, \quad \sin\theta = \frac{y}{r}.$$

Substituting this, we get (2.24). Since $\tan\theta = \frac{y}{x}$, we may define $\arg z$ as $\arctan\frac{y}{x}$ (cf. Exercise 2.34, (ii)).

Recall **de Moivre's formula**

$$(\cos\theta + i\sin\theta)^n = \cos n\theta + i\sin n\theta, \quad n \in \mathbb{Z}. \tag{1.30}$$

Exercise 2.22. Prove (1.30) by induction.

Exercise 2.23. (i) For $z_1, z_2 \neq 0$, prove that

$$\arg(z_1 z_2) = \arg z_1 + \arg z_2$$

and that

$$\arg\left(\frac{z_1}{z_2}\right) = \arg z_1 - \arg z_2.$$

Interpret the results in the light of the exponential law (1.15).
(ii) Deduce from (i) that for $0 \neq z \in \mathbb{C}$ and $\theta \in \mathbb{R}$, $(\cos\theta + i\sin\theta)z$ is the point on the circle $r = |z|$ rotated in the positive direction by θ. What about $(\cos\theta - i\sin\theta)z$?
(iii) Interpret (ii) from the point of view of Euler's identity (1.20). Cf. Remark 1.4.

Exercise 2.24. Define the sequence a_n and b_n by

$$a_n + ib_n = (1+\sqrt{3}i)^n \quad (n \in \mathbb{N}).$$

Find $a_n b_{n-1} - a_{n-1}b_n$ and $a_n a_{n-1} + b_n b_{n-1}$.

Solution. Using the polar form, we may write $a_n + ib_n = (1+\sqrt{3}i)^n = 2^n(\cos\frac{n}{3}\pi + i\sin\frac{n}{3}\pi)$. Hence

$$a_{n-1} + ib_{n-1} = (1+\sqrt{3}i)^{n-1} = 2^{n-1}\left(\cos\frac{n-1}{3}\pi + \sin\frac{n-1}{3}\pi\right).$$

Comparing the coefficients, we get

$$a_n = 2^n\cos\frac{n}{3}\pi, \quad b_n = 2^n\sin\frac{n}{3}\pi,$$

$$a_{n-1} = 2^{n-1}\cos\frac{n-1}{3}\pi, \quad b_{n-1} = 2^{n-1}\sin\frac{n-1}{3}\pi.$$

Therefore we have

$$a_n b_{n-1} - a_{n-1}b_n = 2^{2n-1}\sin\left(\frac{n-1}{3}\pi - \frac{n}{3}\pi\right) = -2^{2n-2}\sqrt{3},$$

$$a_n a_{n-1} + b_n b_{n-1} = 2^{2n-1}\cos\left(\frac{n}{3}\pi - \frac{n-1}{3}\pi\right) = 2^{2n-2}.$$

Another solution (Tricky). Consider

$$(a_{n-1} - ib_{n-1})(a_n + ib_n) = (1-\sqrt{3}i)^{n-1}(1+\sqrt{3}i)^n = 2^{2n-1}(1+\sqrt{3}i).$$

Drill 2.1. Express the following complex numbers in the form $x + iy$ (or in polar form) and plot them on $\mathbb{C}$.
(i) $(3+4i)+(2+7i)$ (ii) $(2+3i)(6+5i)$ (iii) $(4+i)(2+3i)(6+9i)$
(iv) $(1+i)(2-i)(1-i)$ (v) $\frac{1+i}{3-i}$ (vi) $\frac{3-2i}{1+i}$ (vii) $\frac{1+i}{(1-i)^2}$
(viii) $\frac{12+8i}{2-3i} + \frac{52+13i}{13i}$ (ix) $(1+i)^{16}$ (x) $(1-i)^{16}$ (xi) $\left(\frac{34}{(1-4i)(5+3i)}\right)^2$
(xii) $\frac{i}{1+i} + \frac{1+i}{i}$ (xiii) i^{17}.

Drill 2.2. Compute the following absolute values by (2.18).
(i) $|(3+4i)(\sqrt{6}+i)(2-\sqrt{3}i)|$ (ii) $\left|\frac{(1+i)(\sqrt{3}+7i)}{4+6i}\right|$ (iii) $|i(1-i)(2+3i)|$
(iv) $\left|\frac{(6+7i)(4-2i)}{4+2i}\right| \cdot \left|\frac{-1}{7+6i}\right|$.

Example 2.1. We find the fifth root of 2 in $\mathbb{C}$. We are to find z such that $z^5 = 2$. Raising the polar form $z = r(\cos\theta + i\sin\theta)$ of z to the fifth power by (1.29), we obtain

$$z^5 = r^5(\cos 5\theta + i\sin 5\theta),$$

i.e. the polar coordinate is $\begin{pmatrix} r^5 \\ 5\theta \end{pmatrix}$. On the other hand, the polar form of 2 is $2(\cos 0 + i\sin 0)$, i.e. the polar coordinate is $\begin{pmatrix} 2 \\ 0 \end{pmatrix}$, whence we have

$$r^5 = 2,\ 5\theta = 0 + 2\pi n,\ n \in \mathbb{Z}.$$

Hence $r = \sqrt[5]{2}$ and $\theta = \frac{2n}{5}\pi \quad (n \in \mathbb{Z})$.

When we substitute the values of θ back into the polar form, we may restrict the range of n to the least positive residue modulo 5 since sine and cosine are periodic functions of period 2π:

$$\left\{ 5\left(\frac{n}{5} - \left[\frac{n}{5}\right]\right) \middle| n \in \mathbb{Z} \right\} = \{0, 1, 2, 3, 4\},$$

where $[\alpha]$ is the integer part of α in Exercise 2.13. Hence, corresponding to five values of $\theta = 0, \frac{2\pi}{5}, \frac{4\pi}{5}, \frac{6\pi}{5}, \frac{8\pi}{5}$, there are five values of z. For $\theta = 0$, $z = \sqrt[5]{2}$. It is instructive to express other values of z using roots (cf. Exercise 2.27).

Remark 2.2. The procedure in Example 2.1 clearly gives all the solutions to the equation $z^n = 1$, i.e. all n-th roots of 1: $z = e^{\frac{2\pi ik}{n}}$, $k = 0, 1, \ldots, n-1$. Note that those with $\gcd(n, k) = 1$ are **primitive nth roots** of 1.

Also Example 2.1 gives a method for treating an equation of the form $z^n = b \neq 0$. In the first place the radius is $|z| = |b|^{1/n}$ and there remains the determination of the argument.

Exercise 2.25. Express the following complex numbers in the form $x + iy$ and plot them on $\mathbb{C}$.
(i) $\sqrt{i}$ (ii) cube roots of i (iii) 6th roots 1 (iv) 5th roots of -1
(v) $z = \sqrt[5]{-32}$ (vi) cube roots of $1 - i$ (vii) cube roots of $4\sqrt{2} + 4\sqrt{2}i$ (viii) cube roots of $\sqrt{3} - i$ (ix) 4th roots of $-1 + \sqrt{3}i$ (x) 4th roots of i (xi) $z = \sqrt[4]{-1+i}$.

Solution. (i) We are to find z such that $z^2 = i$. Squaring the polar form $z = r(\cos\theta + i\sin\theta)$ of z by (1.30), we obtain

$$z^2 = r^2(\cos 2\theta + i\sin 2\theta),$$

i.e. the polar coordinate (of the corresponding vector $\boldsymbol{z}$) is $\begin{pmatrix} r^2 \\ 2\theta \end{pmatrix}$. On the other hand, the polar form of i is $\cos\frac{\pi}{2} + i\sin\frac{\pi}{2}$, i.e. the polar coordinate is $\begin{pmatrix} 1 \\ \frac{\pi}{2} \end{pmatrix}$, whence we have

$$r^2 = 1, \quad 2\theta = \frac{\pi}{2} + 2n\pi, \quad n \in \mathbb{Z}.$$

Hence $r = 1$ and $\theta = \frac{\pi}{4}, \frac{5\pi}{4}$.

Another solution. Squaring the rectangular coordinates expression $z = x + yi$, we have

$$i = z^2 = (x+yi)^2 = x^2 - y^2 + 2xyi.$$

Comparing the real and imaginary parts, we get the system of equations $x^2 - y^2 = 0$ and $2xy = 1$. This has two solutions $x = y = \frac{\sqrt{2}}{2}$, $x = y = -\frac{\sqrt{2}}{2}$, which give rise to two solutions: $z = \frac{\sqrt{2}}{2} + \frac{\sqrt{2}}{2}i$, $z = -\frac{\sqrt{2}}{2} - \frac{\sqrt{2}}{2}i$, in conformity with the above solutions.

(ii) This can be solved using the polar form. We give another method. Raising $z = x + yi$ to the cube power, we obtain by the binomial theorem

$$i = z^3 = (x+yi)^3 = x^3 - 3xy^2 + (3x^2y - y^3)i$$

which is the same as the system of equations $x^3 - 3xy^2 = 0$, $3x^2y - y^3 = 1$. From this we get the solutions $z = -i$, $z = \frac{\sqrt{3}}{2} + \frac{1}{2}i$, $z = -\frac{\sqrt{3}}{2} - \frac{1}{2}i$.

(vii) Solution is similar to that of (i). The polar form of $4\sqrt{2} + 4\sqrt{2}i$ is $8(\cos\frac{\pi}{4} + i\sin\frac{\pi}{4})$. We are to find z such that $z^3 = 4\sqrt{2} + 4\sqrt{2}i$. Since

$$z^3 = r^3(\cos 3\theta + i\sin 3\theta),$$

we have, comparing the polar coordinates,

$$r^3 = 8, \quad 3\theta = \frac{\pi}{4} + 2n\pi, \quad n \in \mathbb{Z}.$$

Hence $r = 2$ and $\theta = \frac{2n+\frac{1}{4}}{3}\pi$ $(n = 0, 1, 2)$.

(x') We solve the equation

$$z^4 + b^4 = 0, \quad b > 0. \tag{2.25}$$

As is remarked above, $|z| = b$ and the argument θ can be determined from

$$z^4 = -1 = e^{\pi i}.$$

Hence $\theta = \frac{\pi}{4}, \frac{3\pi}{4}, \frac{5\pi}{4}, \frac{7\pi}{4}$ and $z = \pm\frac{1+i}{\sqrt{2}}, \pm\frac{1-i}{\sqrt{2}}$. We note that by the decomposition $z^4 + 1 = z^4 - i^2 = (z^2 - i)(z^2 + i)$, this includes the equation in (i). We also note another decomposition $z^4 + 1 = z^4 + 2z^2 + 1 - (\sqrt{2}z)^2 = (z^2 - \sqrt{2}z + 1)(z^2 + \sqrt{2}z + 1)$ by which we may readily obtain the solutions.

Exercise 2.26. Use (1.30) to deduce the **duplication** and **triplication formulas** for the sine and cosine functions

$$\sin 2\theta = 2\sin\theta\cos\theta, \cos 2\theta = \cos^2\theta - \sin^2\theta = 2\cos^2\theta - 1 = 1 - 2\sin^2\theta,$$

$$\sin 3\theta = 3\sin\theta - 4\sin^3\theta, \quad \cos 3\theta = 4\cos^3\theta - 3\cos\theta.$$

Solution. From de Moivre's formula, we have

$$(\cos\theta + i\sin\theta)^2 = \cos 2\theta + i\sin 2\theta,$$

$$(\cos\theta + i\sin\theta)^3 = \cos 3\theta + i\sin 3\theta.$$

On the other hand, by the binomial theorem

$$(\cos\theta + i\sin\theta)^2 = \cos^2\theta - \sin^2\theta + 2i\sin\theta\cos\theta,$$

$$(\cos\theta + i\sin\theta)^3 = \cos^3\theta - 3\sin^2\theta\cos\theta + i(3\sin\theta\cos^2\theta - \sin^3\theta).$$

Comparing the real and imaginary parts of the above two formulas, we get the duplication formulas

$$\sin 2\theta = 2\sin\theta\cos\theta, \quad \cos 2\theta = \cos\theta - \sin\theta = 2\cos^2\theta - 1 = 1 - 2\sin^2\theta,$$

and triplication formulas

$$\sin 3\theta = 3\sin\theta - 4\sin^3\theta, \quad \cos 3\theta = 4\cos^3\theta - 3\cos\theta.$$

Exercise 2.27. Similarly to Exercise 2.26, deduce the quintuplicate formula

$$\sin 5\theta = 16\sin^5\theta - 20\sin^3\theta + 5\sin\theta,$$

$$\cos 5\theta = 16\cos^5\theta - 20\cos^3\theta + 5\cos\theta.$$

Using the first to find the values $\sin\frac{\pi}{5}$, $\cos\frac{\pi}{5}$.

Solution. In the same way as above, we compare the 5th powers of $e^{i\theta}$ and $\cos 5\theta + i \sin 5\theta$.

To find the value of $\sin\theta$ with $\theta = \frac{\pi}{5}$, we note that $\sin 5\theta = 0$, whence we obtain the quartic equation

$$16 \sin^4 \theta - 20 \sin^2 \theta + 5 = 0$$

whose solution satisfying the restriction $0 < \sin\theta < \sin\frac{\pi}{4} = \frac{1}{\sqrt{2}}$ is $\sin\theta = \left(\frac{10-2\sqrt{5}}{16}\right)^{1/2}$. More interesting is the value of the cosine:

$$2\cos\frac{\pi}{5} = \tau = \frac{1+\sqrt{5}}{2} = 1.618\cdots \tag{2.26}$$

known as the **golden ratio**, cf. e.g. [Chakraborty *et al.* (2009)].

Another solution.

$$\sin 5\theta = \sin(3\theta + 2\theta) = \sin 3\theta \cos 2\theta + \cos 3\theta \sin 2\theta,$$

$$\cos 5\theta = \cos(3\theta + 2\theta) = \cos 3\theta \cos 2\theta - \sin 3\theta \sin 2\theta.$$

Using the results of Exercise 2.26, the conclusion follows.

Exercise 2.28. (n-plication formula) Use (1.30) to deduce the following for $m \in \mathbb{N}$ (i) $\cos mz = P(\cos z)$ (ii) $\sin mz = Q(\sin z)$ (for m odd) and $\sin mz = \cos z Q(\sin z)$ (for m even), where $P(w)$ and $Q(w)$ are polynomials in w.

Solution. We treat the case $n = 2m + 1$ for the sine function only. Comparing the imaginary parts of (1.30), we deduce that

$$\sin(2m+1)\theta = \sum_{k=0}^{m} (-1)^k \binom{2m+1}{2k+1} \sin^{2k+1}\theta \cos^{2(m-k)}\theta$$

$$= \sum_{r=0}^{m} (-1)^{m-r} \binom{2m+1}{2m+1-2r} \sin^{2m+1-2r}\theta (\cos^2\theta)^r$$

on putting $r = m - k$. Hence

$$\sin(2m+1)\theta = \sum_{r=0}^{m} (-1)^{m-r} \binom{2m+1}{2r} \sin^{2m+1-2r}\theta \sum_{s=0}^{r} \binom{r}{s} (-1)^s \sin^{2s}\theta$$

$$= \sum_{r=0}^{m} \sum_{l=0}^{r} (-1)^{m-l} \binom{2m+1}{2r} \binom{r}{r-l} \sin^{2m+1-2l}\theta,$$

where we put $r - s = l$. Hence,

$$\frac{\sin(2m+1)\theta}{\sin\theta} = \sum_{l=0}^{m}(-1)^{m-l}\sum_{r=l}^{m}\binom{2m+1}{2r}\binom{r}{l}\sin^{2m-2l}\theta, \tag{2.27}$$

$$T_n(x) = P(n\arccos x) = 2^{n-1}x^n + \cdots$$

is called the **Čebysëv polynomial of the first kind** of degree n and the companion $U_n(x) = \frac{\sin((n+1)\arccos x)}{\sin(\arccos x)}$ is called **Čebysëv polynomial of the second kind.** For introduction to the theory of Čebysëv polynomials, cf. [Chakraborty *et al.* (2009), pp. 10-22].

Exercise 2.29. Let $2 \le q \in \mathbb{N}$. Suppose $\zeta^q = 1$ but $\zeta^n \ne 1$ for any $1 \le n < q$. Then prove that

$$\prod_{n=1}^{q-1}(1-\zeta^n) = q. \tag{2.28}$$

Solution. Using (2.2), we see that for $z \ne 1$, $\sum\limits_{k=0}^{q-1} z^k = \frac{z^q-1}{z-1}$. Since the right-hand side can be factored as $\prod_{n=1}^{q-1}(z-\zeta^n)$, it follows that

$$\sum_{k=0}^{q-1} z^k = \prod_{n=1}^{q-1}(z-\zeta^n). \tag{2.29}$$

(2.29) with $z = 1$ reduces to (2.28).

Exercise 2.30. Suppose n is a positive odd integer. Use (2.27) to deduce that

$$\frac{\sin n\theta}{\sin\theta} = (-4)^{\frac{n-1}{2}} \prod_{1\le j\le (n-1)/2}\left(\sin^2\theta - \sin^2\frac{2\pi j}{n}\right). \tag{2.30}$$

Solution. From (2.27) it follows that the left-hand side of (2.30) is a polynomial in $X^2 = \sin^2\theta$ of degree $\frac{n-1}{2}$. Then the zeros of the right-hand side are $X = \sin\frac{2\pi j}{n}$, $-\frac{n}{2} \le j < \frac{n}{2}, j \ne 0$. It remains to evaluate the coefficient of the degree $\frac{n-1}{2}$ term. Use will be made of the combinatorial identity

$$2^{n-1} = \sum_{j=0}^{[\frac{n-1}{2}]}\binom{n}{2j}.$$

Exercise 2.31. Suppose n is a positive odd integer. Use Exercise 2.30 and (2.27) to deduce that

$$\prod_{j=1}^{n-1}\left(2\sin\frac{\pi j}{n}\right) = n. \tag{2.31}$$

This is true for any $n \in \mathbb{N}$. Cf. Exercise 2.32.

Solution. This follows on comparing the constant terms of those two identities, the constant term of (2.27) being $\prod_{j=1}^{\frac{n-1}{2}} \left(2\sin\frac{2\pi j}{n}\right)^2$.

Exercise 2.32. Deduce (2.31) from Exercise 2.29.

Solution. We may put $\zeta = e^{\frac{2\pi i}{n}}$ in (2.28). Then noting that

$$(1-\zeta^r)(1-\zeta^{-r}) = \left(2\frac{\zeta^{r/2}-\zeta^{-r/2}}{2i}\right)^2 = \left(2\sin\frac{\pi r}{n}\right)^2$$

and that $\prod_{r=1}^{n-1}(1-\zeta^r)(1-\zeta^{-r}) = \prod_{r=1}^{n-1}(1-\zeta^r)^2$, we conclude that

$$n^2 = \left(\prod_{r=1}^{n-1}(1-\zeta^r)\right)^2 = \left(\prod_{r=1}^{n-1} 2\sin\frac{\pi r}{n}\right)^2.$$

Hence (2.31) follows.

By Corollary 1.4 to Theorem 1.9 (= Theorem 2.22), any polynomial $f(x) \in \mathbb{C}[x]$ can be decomposed into linear factors, or in other words, all roots of $f(x) = 0$ lie in $\mathbb{C}$. Hence $\mathbb{C}$ is **an algebraically closed field**.

So far we have introduced the following number systems and beyond $\mathbb{C}$ we cannot go further:

$$\mathbb{N} \subset \mathbb{Z} \subset \mathbb{Q} \subset \mathbb{R} \subset \mathbb{C}. \tag{2.32}$$

Hereafter we shall freely use the set-theoretic notation and two logical symbols: "$\forall \sim$" to mean "for any $\sim$" and "$\exists \sim$" s.t. $\cdots$ to mean "there exists $\sim$ such that $\cdots$."

Review Problem. For z_1, z_2 plot the following on $\mathbb{C}$: $z_1 \pm z_2$, $z_1 z_2$, z_1/z_2 $(z_2 \neq 0)$.

2.3 Power series again

(1.11) is an almost perfect tool for finding the radius r of convergence of a power series as long as the limit exists, which is most often the case. A universal formula for r is given in Theorem 2.3 below due to Cauchy. To state it, we recall the notion of limes principals. If you're familiar with it, you can skip the next subsection.

2.3.1 *Limes principals*

We shall expound the two defining conditions for limes principals. Let $\{a_n\} = \{a_n\}_{n=1}^{\infty}$ be a real sequence, which we abbreviate as $\{a_n\} \subset \mathbb{R}$, by interpreting $\{a_n\}$ to mean the set $\{a_n | n \in \mathbb{N}\}$ of all terms of the sequence. The set $\{a_n\}$ may be finite.

We adopt the monotone convergence theorem (MCT) and Weierstrass theorem as our axioms for continuity of real numbers.

Since as can be seen $\lim_{n\to\infty} \inf a_n = -\lim_{n\to\infty} \sup(-a_n)$, we shall confine ourselves to the case of lim sup a_n.

First there are two cases to consider:

(i) $\{a_n\}$ is not upward bounded, i.e. for any large $M > 0$, there exists an $n_0 = n_0(M) \in \mathbb{N}$ such that $a_{n_0} > M$. In this case we define

$$\lim_{n\to\infty} \sup a_n = \infty. \tag{2.33}$$

(ii) $\{a_n\}$ is upward bounded. Then each subset

$$A_n = \{a_n, a_{n+1}, \dots\}$$

of $\{a_n | n \in \mathbb{N}\}$ has a finite supremum: $\sup A_n$, in view of Weierstrass' theorem.

Now the new sequence $\sup A_n$ is a monotone non-increasing sequence:

$$\sup A_{n+1} \leq \sup A_n.$$

There are two possible cases: the set $\{A_n | n \in \mathbb{N}\}$ is downward bounded or not. If it is downward bounded, then the MCT implies that the limit exists and is equal to the inf of that set. Then we define

$$\lim_{n\to\infty} \sup a_n = \inf\{\sup A_n\}. \tag{2.34}$$

If it is not downward bounded, then $\inf\{\sup A_n | n \in \mathbb{N}\} = -\infty$ and we define

$$\limsup a_n = -\infty. \tag{2.35}$$

For applications, the following characterization is effective.

Theorem 2.2. *If the* $\limsup A$ *of* $\{a_n\}$ *if finite, then it is the number satisfying two conditions*

(i) *Any number* $> A$, *say* $A + \varepsilon, (\forall \varepsilon > 0)$ *is an upper bound for all* a_n *for some* $n_0 = n_0(\varepsilon)$ *onwards:*

$$a_n < A + \varepsilon, \quad n \geq n_0.$$

(ii) *No number* $< A$, *say* $A - \varepsilon, (\forall \varepsilon > 0)$, *can be a lower bound for* $\{\sup A_n\}$: *There exists an* $n_0 = n_0(\varepsilon) \in \mathbb{N}$ *such that*

$$A - \varepsilon < \sup A_{n_0}.$$

Proof. It suffices to consider the case $A = \limsup a_n$ being finite. In this case, we have, for any $\varepsilon > 0$, there exists an n_0 such that

(i)' $\sup A_{n_0} < A + \varepsilon$,

and

(ii)' $A \le \sup A_n$, $\forall n \in \mathbb{N}$.

Since $\sup A_n \le \sup A_{n_0}$ for $n \ge n_0$, it follows that $a_n \le \sup A_n \le \sup A_{n_0} < A + \varepsilon$, whence (i) follows.

By the definition of sup, for any $\varepsilon > 0$, $\sup A_n - \varepsilon$ cannot be an upper bound for $\{A_n\} : \exists \quad n_0 = n_0(\varepsilon) \in \mathbb{N}$ such that

$$A - \varepsilon \le \sup A_n - \varepsilon < a_{n_0}, \quad n \ge n_0,$$

whence (ii).

Conversely, assume (i) and (ii). Then (i) implies that $\sup A_{n_0} \le A + \varepsilon$, i.e. (i)'. That $A > \sup A_{n_0}, for some\ n_0 \in \mathbb{N}$ means that A is not a lower bound for the set $\{A_n | n \in \mathbb{N}\}$, a contradiction. Hence (ii)' holds. This completes the proof. □

2.3.2 *Radius of convergence*

Theorem 2.3. (Cauchy) *The radius of convergence of the power series*

$$\sum_{n=1}^{\infty} a_n z^n \tag{2.36}$$

is given by

$$\frac{1}{\limsup_{n \to \infty} |a_n|^{1/n}}. \tag{2.37}$$

We omit the proof which depends upon the Cauchy criterion.

Example 2.2. ([Chakraborty *et al.* (2009), Exercise 2.1, p. 25]) The radius of convergence of the power series (1.13) for the exponential function is ∞, i.e. it is absolutely and uniformly convergent over the whole plane. We use Theorem 2.3 to conclude this, on appealing to the **Stirling formula** for the gamma function.

$$\Gamma(n+1) = n! \sim \sqrt{2\pi n}\left(\frac{n}{e}\right)^n. \tag{2.38}$$

Since $(n!)^{1/n} \sim (2\pi n)^{1/(2n)} \frac{n}{e} \to \infty$ as $n \to \infty$, the assertion follows.

Example 2.3. (Lambert series, [Kanemitsu and Tsukada (2014)]) Let $\{a_n\} \subset \mathbb{C}$ be such that

$$\limsup |a_n|^{1/n} \leq 1. \tag{2.39}$$

Then by Theorem 2.3, the power series $\sum_{n=1}^{\infty} a_n z^n$ is absolutely convergent in $|z| < 1$. Then the **Lambert series**

$$f(z) = \sum_{n=1}^{\infty} a_n \frac{z^n}{1 - z^n} \tag{2.40}$$

is absolutely convergent in $|z| < 1$ and represents an analytic function. This is because the Majorant series is bounded by $O(\sum_{n=1}^{\infty} |a_n||z|^n)$.

Moreover the power series of this function can be obtained by formal rearrangement of (2.40), i.e.

$$f(z) = \sum_{n=1}^{\infty} b_n z^n \quad (|z| < 1), \tag{2.41}$$

where

$$b_n = \sum_{d|n} a_d, \tag{2.42}$$

where the symbol $d \mid n$ means that d runs through all (positive) divisor of n.

2.3.3 *Function series*

Theorem 2.4. *A uniformly convergent series of analytic functions may be integrated term by term along any curve inside the region of uniform convergence. Namely, if the functions*

$$f_1(z),\ f_2(z),\ \ldots$$

are analytic in D and the series

$$\sum_{n=1}^{\infty} f_n(z) = f(z)$$

is uniformly convergent in D, then for any curve $C \subset D$, we have

$$\int_C f(z)\,\mathrm{d}z = \int_C \sum_{n=1}^{\infty} f_n(z)\,\mathrm{d}z = \sum_{n=1}^{\infty} \int_C f_n(z)\,\mathrm{d}z.$$

Proof. $f_n(z)$ need not be analytic but enough to be continuous in D (since analyticity $\Rightarrow$ continuity, the assumption is excessive). Since $f(z)$ is continuous in D, it follows that the integral $\int_C f_n(z)\,\mathrm{d}z,\ n \in \mathbb{N}$ exists. So does the integral $\int_C s_n(z)\,\mathrm{d}z$ for $s_n(z) = \sum_{i=1}^{n} f_i(z)$. Since $s_n(z)$ converges to $f(z)$ uniformly on D, we have

$$\forall \varepsilon > 0,\ \exists n_0 = n_0(\varepsilon) \in \mathbb{N} \quad s.t. \quad n > n_0 \Rightarrow |s_n(z) - f(z)| < \varepsilon, \quad \forall z \in D.$$

Hence for $n \geq n_0$, we have

$$\left| \int_C (s_n(z) - f(z))\,\mathrm{d}z \right| < \varepsilon\, \Lambda(C),$$

where $\Lambda(C)$ is the length of C so that

$$\lim_{n\to\infty} \int_C s_n(z)\,\mathrm{d}z = \int_C f(z)\,\mathrm{d}z$$

whose left-hand side is nothing other than the definition of

$$\sum_{n=1}^{\infty} \int_C f_n(z)\,\mathrm{d}z.$$

□

Theorem 2.5. *The limit of the uniformly convergent series of analytic functions is interchangeable with integration along any curve lying in the region of its uniform convergence. Namely, if*

$$f_1(z),\ f_2(z),\ \ldots$$

are analytic in D and

$$\lim_{n\to\infty} f_n(z) = f(z),$$

uniformly in D, then for any Jordan curve $C \subset D$, we have

$$\int_C f(z)\,\mathrm{d}z = \lim_{n\to\infty} \int_C f_n(z)\,\mathrm{d}z = \int_C \lim_{n\to\infty} f_n(z)\,\mathrm{d}z.$$

Definition 2.3. If a sequence (respectively, series) of functions defined on D are uniformly convergent on any bounded closed subset of D (i.e., on any compact subset D' such that $D' \subset D$), we say that the sequence (respectively, series) is **uniformly convergent** on D **in the wide sense**.

Theorem 2.6. *If the functions*

$$f_1(z),\ f_2(z),\ \ldots$$

are (i) *analytic in D and* (ii) *the series*

$$\sum_{n=1}^{\infty} f_n(z)$$

is uniformly convergent in D in the wide sense, then its sum

$$f(z) = \sum_{n=1}^{\infty} f_n(z)$$

is analytic in D and its derivative may be obtained by termwise differentiation:

$$f'(z) = \sum_{n=1}^{\infty} f_n'(z).$$

Also, the termwise differentiated series is uniformly convergent in the wide sense in D.

Corollary 2.1. *Any function series*

$$\sum_{n=1}^{\infty} f_n(z) := f(z)$$

that is uniformly convergent in the wide sense in D is termwise differentiable infinitely many times:

$$f^{(k)}(z) = \sum_{n=1}^{\infty} f_n^{(k)}(z), \quad k \in \mathbb{N}$$

(and the k-times differentiated series is also uniformly convergent in the wide sense in D).

Proof. We shall prove both Theorem 2.6 and its Corollary 2.1 at the same time. By the Cauchy integral formula in Theorem 1.9 (= Theorem 2.22), we have for any rectifiable simple curve C in D and any point z in C,

$$f(z) = \sum_{n=1}^{\infty} f_n(z) = \frac{1}{2\pi i} \sum_{n=1}^{\infty} \int_C \frac{f_n(w)}{w-z}\, dw.$$

But $\sum_{n=1}^{\infty} f_n(w)$ is uniformly convergent on C, and so Theorem 2.4 allows us to integrate term by term after multiplying $(w-z)^{-1}$:

$$f(z) = \frac{1}{2\pi i} \int_C \sum_{n=1}^{\infty} \frac{f_n(w)}{w-z}\, dw = \frac{1}{2\pi i} \int_C \frac{f(w)}{w-z}\, dw.$$

Hence the Cauchy integral formula holds for $f(z)$ and so it follows that $f(z)$ is analytic in C and that

$$f^{(k)}(z) = \frac{k!}{2\pi i}\int_C \frac{f(w)}{(w-z)^{k+1}}\,\mathrm{d}w, \quad k \in \mathbb{N}\cup\{0\}.$$

Let

$$S_n(z) = \sum_{k=1}^{n} f_k(z)$$

be the n-th partial sum of

$$\sum_{n=1}^{\infty} f_n(z)$$

and take any bounded closed subset D' in D. Then take any simple closed contour $C \subset D$ of finite length $\Lambda(C)$ containing D' and suppose $dist\ (D', C) = \delta > 0$. Then we have

$$S_n^{(k)}(z) = \frac{k!}{2\pi i}\int_C \frac{S_n(w)}{(w-z)^{k+1}}\,\mathrm{d}w, \quad k \in \mathbb{N}\cup\{0\}.$$

Hence, it follows that

$$\left|f^{(k)}(z) - S_n^{(k)}(z)\right| \le \frac{k!}{2\pi\delta^{k+1}}\,\Lambda(C)\,\max_{w\in C}|f(w) - S_n(w)|,$$

whence we have

$$\forall\varepsilon > 0, \quad \exists n_0 = n_0(\varepsilon) \in N \ s.t.\ n > n_0 \ \Rightarrow$$

$$|f(w) - S_n(w)| < \varepsilon$$

on C. Hence,

$$\lim_{n\to\infty} S_n^{(k)}(z) = f^{(k)}(z)$$

uniformly on D'. □

The following corollary is the genesis of Theorem 1.1.

Corollary 2.2. (Weierstrass double series theorem) *Suppose* $\{f_n(z)\}$ *are analytic in* $|z - z_0| < r$ *and has the Taylor expansion*

$$f_n(z) = \sum_{k=0}^{\infty} a_k^{(n)}(z - z_0)^k.$$

Then if

$$\sum_{n=1}^{\infty} f_n(z) = f(z)$$

uniformly in the wide sense on $|z - z_0| < r$*, then* $f(z)$ *is analytic on* $|z - z_0| < r$ *and its Taylor expansion is given by*

$$f(z) = \sum_{k=0}^{\infty} a_k (z - z_0)^k \quad (|z - z_0| < r), \quad a_k = \sum_{n=1}^{\infty} a_k^{(n)}.$$

I.e. the iterates of the double series coincide – the order of summation being interchangeable –

$$\sum_{n=1}^{\infty}\sum_{k=0}^{\infty} a_k^{(n)} (z - z_0)^k \left(= \sum_{n=1}^{\infty} f_n(z) = f(z) \right) = \sum_{k=0}^{\infty} a_k (z - z_0)^k.$$

Proof. This is a special case of Theorem 2.5. The relation between coefficients follows from Theorem 2.27:

$$a_k = \frac{1}{k!} f^{(k)}(z_0) = \frac{1}{k!} \sum_{n=1}^{\infty} f_n^{(k)}(z_0) = \sum_{n=1}^{\infty} a_k^{(n)}.$$

□

Theorem 2.7. *If* $\{f_n(z)\}$ *are analytic in* D *and*

$$\lim_{n\to\infty} f_n(z) = f(z)$$

uniformly in the wide sense on D*, then* $f(z)$ *is again analytic on* D *and*

$$f^{(k)}(z) = \lim_{n\to\infty} f_n^{(k)}(z)$$

uniformly in D *in the wide sense.*

2.4 Improper integrals*

For a real-valued function $f(x)$ defined and bounded on the finite interval $[a, b]$, the important quantity (1-dimensional) definite integral $\int_a^b f(x)\,dx$ is defined in Definition A.1. In this subsection we are going to generalize this notion of definite integrals to

Case (i) the interval is infinite, e.g. of the type $[a, \infty)$

and

Case (ii) the function $f(x)$ is unbounded at an end-point of the interval, e.g. $\lim_{\varepsilon\to+0} |f(a + \varepsilon)| = \infty$

and show that in these cases we may also define a quantity corresponding to the area with sign of the figure formed by the graph of the function $y = f(x)$ and the x-axis.

Infinite integrals are analogous to infinite series and $I(b) = \int_0^b a(t)\,\mathrm{d}t$ corresponds to the partial sum $S_n = \sum_{m=0}^{n} a(m)$, so that $I(b)$ is what we call, the "partial integral". Indeed, from the point of view of Stieltjes integrals, both of them are special cases of Stieltjes integrals (cf. [Apostol (1957)]. Although b varies continuously $\to \infty$ and n discretely $\to \infty$, those theorems on infinite series have their counterparts in infinite integrals. By stating Case (ii) rather in detail, we hope that the reader will be able to perceive the corresponding results on infinite series, and *a fortiori* the limits of a sequence since the limit of a sequence is the same as the limit of the corresponding telescoping series.

Case (i) We define the **infinite integral** or **improper integral of the first kind** $\int_a^\infty f(x)\,\mathrm{d}x$ as the limit $b \to \infty$ of the proper integral $I(b) = \int_a^b f(x)\,\mathrm{d}x$:

$$\int_a^\infty f(x)\,\mathrm{d}x = \lim_{b\to\infty} \int_a^b f(x)\,\mathrm{d}x.$$

Similarly, we define

$$\int_{-\infty}^b f(x)\,\mathrm{d}x = \lim_{a\to-\infty} \int_a^b f(x)\,\mathrm{d}x.$$

If for some constant c, both $\int_{-\infty}^c f(x)\,\mathrm{d}x$ and $\int_c^\infty f(x)\,\mathrm{d}x$ exist, then we denote their sum by $\int_{-\infty}^\infty f(x)\,\mathrm{d}x$:

$$\begin{aligned}\int_{-\infty}^\infty f(x)\,\mathrm{d}x &= \int_{-\infty}^c f(x)\,\mathrm{d}x + \int_c^\infty f(x)\,\mathrm{d}x \\ &= \lim_{a\to\infty} \int_{-a}^c f(x)\,\mathrm{d}x + \lim_{b\to\infty} \int_c^b f(x)\,\mathrm{d}x,\end{aligned}$$

where it is important that a and b independently $a, b \to \infty$. Especially, if $a = b \to \infty$, i.e. $\lim\limits_{a\to\infty} \int_{-a}^a f(x)\,\mathrm{d}x$ is called the **Cauchy principal value** and denoted

$$PV \int_{-\infty}^\infty f(x)\,\mathrm{d}x.$$

If $\int_{-\infty}^\infty f(x)\,\mathrm{d}x$ exists, then $PV \int_{-\infty}^\infty f(x)\,\mathrm{d}x$ necessarily exists. But the converse does not hold as the example $f(x) = x^3$ shows.

In the case of (ii), we write $\int_{a+0}^b f(x)\,\mathrm{d}x$ or simply $\int_a^b f(x)\,\mathrm{d}x$ and call it an **improper integral of the second kind**. This is defined as the limit as $\varepsilon \to +0$ of the integral $\int_{a+\varepsilon}^b f(x)\,\mathrm{d}x$:

$$\int_{a+0}^b f(x)\,\mathrm{d}x = \lim_{\varepsilon\to+0} \int_{a+\varepsilon}^b f(x)\,\mathrm{d}x.$$

This, however, reduces to (i) by the change of variable $x = a + \frac{1}{t}$, $t > 0$. If the function $f(x)$ is unbounded at the right end-point b, then we define

$$\int_a^{b-0} f(x)\,dx = \lim_{\varepsilon\to+0}\int_a^{b-\varepsilon} f(x)\,dx,$$

but this also reduces to the case (i) by the change of variable $x = b - \frac{1}{t}$, $t > 0$.

If a primitive function can be easily found, then we may apply Corollary 2.4 and take the limit. E.g.

$$\begin{aligned}\int_{0+}^{1} \log x\,dx &= \lim_{\varepsilon\to+0}\,[x\log x - x]_\varepsilon^1 = \lim_{\varepsilon\to+0}\frac{\log\varepsilon}{1/\varepsilon} - 1\\ &= \lim_{\varepsilon\to+0}\frac{1/\varepsilon}{-\left(1/\varepsilon^2\right)} - 1 = -1, \qquad (2.43)\end{aligned}$$

where in the penultimate step we applied L'Hospital's rule.

Similarly,

$$\int_1^\infty x^{-r}\,dx = \begin{cases}\lim\limits_{b\to\infty}\left[\frac{x^{1-r}}{1-r}\right]_1^b, & r\neq 1\\ \lim\limits_{b\to\infty}\,[\log x]_1^b, & r = 1\end{cases} = \begin{cases}\dfrac{-1}{1-r}, & r > 1\\ \infty, & r\le 1,\end{cases}$$

and

$$\int_{0+}^{1} x^{-r}\,dx = \begin{cases}\dfrac{1}{1-r}, & r < 1\\ \infty, & r \ge 1.\end{cases}$$

These shift to each other under $x = 1/t$.

Instead of series with positive terms, we take infinite integrals of positive-valued functions $f(x)$. Then the comparison test reads: for positive-valued functions $f(x)$, $g(x)$, $f(x) \le g(x)$ for $a \le x$,

$$\int_a^\infty g(x)\,dx < \infty \Rightarrow \int_a^\infty f(x)\,dx < \infty.$$

Hence by comparing with $\int_1^\infty x^{-r}\,dx$, which is an analogue of the sum of negative powers of natural numbers $\sum\limits_{n=1}^{\infty} n^{-r}$, we may verify the convergence.

Also if $\lim\limits_{x\to\infty}\frac{f(x)}{g(x)} = 0$ and $\int_a^\infty g(x)\,dx < \infty$, then we have $\int_a^\infty f(x)\,dx < \infty$.

For a complex-valued function $f(x)$ we consider its modulus and apply an analogue of the Weierstrass M-test:

Theorem 2.8. (Weierstrass M-test) *For the function $f(x,y)$, $x \in [a,b]$ of two variables $[$or on $(a,b)]$ and $y \in Y$, there exists a positive-valued function $M(x)$ such that for $\forall y \in Y$, we have $|f(x,y)| \le M(x)$ and $\int_a^\infty M(x)\,\mathrm{d}x < \infty$ [or $\int_a^b M(x)\,\mathrm{d}x < \infty$], then $\int_a^\infty f(x,y)\,\mathrm{d}x$ [or $\int_a^b f(x,y)\,\mathrm{d}x$] converges absolutely and uniformly on Y.*

In the case of real functions, convergence is a delicate problem and there are some theorems known assuring termwise differentiation and the following is often used.

Theorem 2.9. [Apostol (1957), Theorem 13-14, p. 403] *Assume that each f_k is a real-valued function defined and differentiable at each $x \in (a,b)$. Assume that for at least one point $x_0 \in (a,b)$, the series $\sum_{k=1}^\infty f_k(x_0)$ is convergent. Assume further that there exists a function g such that $\sum_{k=1}^\infty f_k'(x) = g(x)$ uniformly on (a,b). Then there exists a function f such that $\sum_{k=1}^\infty f_k(x) = f(x)$ uniformly on (a,b) and that if $x \in (a,b)$, then the derivative $f'(x)$ exists and equals $\sum_{k=1}^\infty f_k'(x) = f'(x)$.*

For differentiation under integral sign, we have

Theorem 2.10. *Suppose that $f(x,t)$ is integrable in x on $[a,b]$ for all t near the point α and that the limit function $\lim_{t\to\alpha} f(x,t) = f(x,\alpha)$ is integrable on $[a,b]$. Further suppose that the limit is uniform in $x \in [a,b]$. Then we may take the limit under the integral sign:*

$$\lim_{t\to\alpha} \int_a^b f(x,t)\,\mathrm{d}x = \int_a^b \lim_{t\to\alpha} f(x,t)\,\mathrm{d}x. \tag{2.44}$$

Proof. By the uniformity of convergence, given $\varepsilon > 0$, there exists a $\delta = \delta(\varepsilon) > 0$ such that for $0 < |t-\alpha| < \delta$,

$$|f(x,t) - f(x,\alpha)| < \varepsilon \tag{2.45}$$

for all $x \in [a,b]$. Hence

$$\left| \int_a^b f(x,t)\,\mathrm{d}x - \int_a^b f(x,\alpha)\,\mathrm{d}x \right| \le \varepsilon \int_a^b \mathrm{d}x = (b-a)\varepsilon,$$

which completes the proof. □

Corollary 2.3. *Suppose that $f(x,t)$ is integrable in x on $[a,b]$ for all t in a certain domain T, that $f(x,t)$ is differentiable in $t \in T$ for every $x \in [a,b]$ and that $f_t = \frac{\partial f}{\partial t}$ is continuous in both the variables x and t. Then we may differentiate under the integral sign:*

$$\frac{\mathrm{d}}{\mathrm{d}t} \int_a^b f(x,t)\,\mathrm{d}x = \int_a^b \frac{\partial}{\partial t} f(x,t)\,\mathrm{d}x. \tag{2.46}$$

Proof. By Theorem 2.10, it suffices to prove that

$$\lim_{h\to 0} \frac{\Delta f}{h} = f_t(x,t)$$

uniformly in $x \in [a,b]$, where $\Delta f = f(x,t+h) - f(x,t)$. By the mean value theorem, there exists a θ, $0 \le \theta \le 1$ such that

$$\Delta f = f(x,t+h) - f(x,t) = f_t(x,t+\theta h)h.$$

Since $f_t(x,t)$ is continuous in $x \in [a,b]$, it is uniformly continuous on $[a,b]$. Hence if $0 < |h|$ is small enough, then

$$\left|\frac{\Delta f}{h} - f_t(x,t)\right| = |f_t(x,t+\theta h) - f_t(x,t)|$$

is arbitrarily small uniformly in $x \in [a,b]$. Hence the convergence is uniform, and the proof is complete.

□

Example 2.4. We find the value of the probability integral, cf. Example 1.30 and Exercise A.6.

$$\int_{-\infty}^{\infty} e^{-t^2}\,\mathrm{d}t = \sqrt{\pi}. \tag{2.47}$$

The integral exists by the comparison test. Consider

$$f(x) = \left(\int_0^x e^{-t^2}\,\mathrm{d}t\right)^2.$$

Then $f'(x) = 2e^{-x^2}\int_0^x e^{-t^2}\,\mathrm{d}t$. Along with f consider

$$g(x) = \int_0^1 \frac{e^{-x^2(t^2+1)}}{t^2+1}\,\mathrm{d}t.$$

Differentiating under the integral sign, we obtain

$$\begin{aligned} g'(x) &= \frac{\mathrm{d}}{\mathrm{d}x}\int_0^1 \frac{1}{t^2+1}e^{-x^2(t^2+1)}\,\mathrm{d}t \\ &= \int_0^1 \frac{1}{t^2+1}\frac{\partial}{\partial x}e^{-x^2(t^2+1)}\,\mathrm{d}t = -2x\int_0^1 e^{-x^2(t^2+1)}\,\mathrm{d}t, \end{aligned}$$

which becomes, by the change of variable $xt = u$, $g'(x) = -2e^{-x^2}\int_0^x e^{-t^2}\,\mathrm{d}t$. Hence it follows that $f'(x) + g'(x) = 0$. We conclude from the Newton-Leibnitz rule (Corollary 1.5) that $f(x) + g(x) = C$. The value of the constant C may be found by putting $x = 0$: $f(0) = 0$, $g(0) = \int_0^1 \frac{1}{t^2+1}\,\mathrm{d}t = \arctan 1 = \frac{\pi}{4}$, i.e. $C = \frac{\pi}{4}$. Hence $f(x) + g(x) = \frac{\pi}{4}$. Letting $x \to \infty$, we conclude that $f(\infty) + g(\infty) = \frac{\pi}{4}$, provided that the limits exist. $f(\infty)$ exists and by taking the limit under the integral sign, we obtain $g(\infty) = 0$. Hence $f(\infty) = \frac{\pi}{4}$ and (2.47) follows.

Example 2.5. (i) (Euler)

$$\int_0^\infty \frac{\log x}{x^2+1}\, dx = 0.$$

This is the sum of improper integrals of the first and the second kind. By the change of variable $x = {}^1/_t$, we have $\int_{0+}^1 \frac{\log x}{x^2+1}\, dx = -\int_1^\infty \frac{\log t}{t^2+1}\, dt$. Hence it suffices to show that $\int_1^\infty \frac{\log t}{t^2+1}\, dt < \infty$. By L'Hospital's rule, we obtain

$$0 \le \lim_{x\to\infty} \frac{\log x}{x^2+1} x^{3/2} \le \lim_{x\to\infty} \frac{\log x}{\sqrt{x}} = \lim_{x\to\infty} \frac{\frac{1}{x}}{\frac{1}{2\sqrt{x}}} = \lim_{x\to\infty} \frac{2}{\sqrt{x}} = 0.$$

Hence by $\int_1^\infty x^{-3/2}\, dx < \infty$ and the comparison theorem, the integral is convergent. Hence

$$\int_0^\infty \frac{\log x}{x^2+1}\, dx = \int_0^1 + \int_1^\infty = -\int_1^\infty \frac{\log x}{x^2+1}\, dx + \int_1^\infty \frac{\log x}{x^2+1}\, dx = 0.$$

We note that such a computation is not possible if convergence is not assured.

(ii)

$$\int_0^\infty t^{s-1} e^{-t}\, dt$$

is absolutely and uniformly convergent for $\sigma = \Re s > 0$ and defines an analytic function, called the **gamma function** $\Gamma(s)$. This is also the sum of improper integrals of the first and the second kind. The modulus of the integrand being $t^{\sigma-1}e^{-t}$, we have

$$\Gamma(\sigma) = \int_0^\infty t^{\sigma-1} e^{-t}\, dt = \int_{0+}^1 t^{\sigma-1} e^{-t}\, dt + \int_1^\infty t^{\sigma-1} e^{-t}\, dt.$$

With $Y \subset [c, d]$, we use the **Majorants**

$$t^{\sigma-1}e^{-t} \le t^{c-1}e^{-t} \quad (0 < t < 1), \quad t^{\sigma-1}e^{-t} \le t^{d-1}e^{-t} \quad (1 \le t)$$

and note that

$$\int_1^\infty t^{d-1} e^{-t}\, dt = O\left(\int_1^\infty t^{-2}\, dt\right) = O(1).$$

2.5 Differentiation*

Let a vector-valued function $\boldsymbol{w} = \boldsymbol{f}(\boldsymbol{z})$ in a vector argument $\boldsymbol{z} = \begin{pmatrix} x \\ y \end{pmatrix} \in D \subset \mathbb{R}^2$ be given by

$$\begin{pmatrix} u \\ v \end{pmatrix} = \boldsymbol{w} = \boldsymbol{f}(\boldsymbol{z}) = \begin{pmatrix} u(\boldsymbol{z}) \\ v(\boldsymbol{z}) \end{pmatrix} = \begin{pmatrix} u(x,y) \\ v(x,y) \end{pmatrix}, \quad \boldsymbol{z} = \begin{pmatrix} x \\ y \end{pmatrix} \in D \subset \mathbb{R}^2, \tag{2.48}$$

which is equivalent to the system of equations

$$\begin{cases} u = u(\boldsymbol{z}) = u(x,y) \\ v = v(\boldsymbol{z}) = v(x,y). \end{cases} \tag{2.49}$$

Definition 2.4. The vector-valued function $\boldsymbol{w} = \boldsymbol{f}(\boldsymbol{z})$ is said to be totally differentiable (or Frechet differentiable) at $\boldsymbol{z}$ if

$$\boldsymbol{f}(\boldsymbol{z}+\boldsymbol{h}) = \boldsymbol{f}(\boldsymbol{z}) + A\boldsymbol{h} + o(|\boldsymbol{h}|) \tag{2.50}$$

as $\boldsymbol{h} \to \boldsymbol{o}$, i.e. $|\boldsymbol{h}| \to 0$. Here A is a matrix of degree 2 and is called the gradient (or the Jacobi matrix) of $\boldsymbol{f}$, denoted $\nabla \boldsymbol{f}$.

We remark that Definition 2.4 entails all the other definitions of derivatives as special cases most of which have division counterpart. However (2.50) **cannot in general have a division counterpart.**

First, if $\boldsymbol{w} = \boldsymbol{f}(\boldsymbol{z})$ is a scalar function $w = f(z)$, then it amounts to

$$f(z+h) = f(z) + f'(z)h + o(h), \tag{2.51}$$

as $h \to 0$ (where by convention we often omit the absolute value sign), which is however equivalent to the usual definition of the differential coefficient of a function of one real or complex variable

$$f'(z) = \lim_{h\to 0} \frac{f(z+h) - f(z)}{h}. \tag{2.52}$$

Secondly, if the variable is a scalar, then we denote it by t and the vector-valued function $\boldsymbol{w} = \boldsymbol{f}(t)$ is a position vector, often denoted $\boldsymbol{x}(t)$ indicating the position of the particle at time t.

$$\boldsymbol{f}(t+h) = \boldsymbol{f}(t) + \frac{\mathrm{d}}{\mathrm{d}t}\boldsymbol{f}(t)h + o(h) \tag{2.53}$$

which has a **division counterpart**

$$\frac{\mathrm{d}}{\mathrm{d}t}\boldsymbol{f}(t) = \boldsymbol{f}'(t) = \lim_{h\to 0} \frac{\boldsymbol{f}(t+h) - \boldsymbol{f}(t)}{h}, \tag{2.54}$$

which corresponds to the velocity vector $\boldsymbol{v}(t) = \frac{\mathrm{d}}{\mathrm{d}t}\boldsymbol{f}(t)$ the instantaneous ratio of change of the position vector. Its derivative is the acceleration $\boldsymbol{a}(t) = \frac{\mathrm{d}}{\mathrm{d}t}\boldsymbol{v}(t) = \frac{\mathrm{d}^2}{\mathrm{d}t^2}\boldsymbol{f}(t)$, the instantaneous ratio of change of the velocity vector.

Thirdly, if $v = 0$ and $\boldsymbol{h} = \begin{pmatrix} h \\ 0 \end{pmatrix}$, then (2.50) entails

$$u(x+h, y) = u(x, y) + \frac{\partial u}{\partial x}h + o(h), \tag{2.55}$$

where $\frac{\partial u}{\partial x}$ is the partial derivative of u with respect to x, i.e. the instantaneous ratio of change of u in x with other variables remaining constant.

It has a division counterpart

$$\frac{\partial u}{\partial x} = u_x = \lim_{h \to 0} \frac{u(x+h, y) - u(x, y)}{h}, \tag{2.56}$$

and similarly,

$$\frac{\partial u}{\partial y} = u_y = \lim_{k \to 0} \frac{u(x, y+k) - u(x, y)}{k}. \tag{2.57}$$

Fourthly, if $v = 0$, then (2.50) entails

$$u(\boldsymbol{z} + \boldsymbol{h}) = u(\boldsymbol{z}) + (\nabla u)\boldsymbol{h} + o(|\boldsymbol{h}|), \tag{2.58}$$

where

$$\nabla u = \left(\frac{\partial u}{\partial x}, \frac{\partial u}{\partial y}\right) = (u_x, u_y). \tag{2.59}$$

Theorem 2.11. *If $\boldsymbol{w} = \boldsymbol{f}(\boldsymbol{z})$ is differentiable at $\boldsymbol{z}$, then it is continuous there and we have*

$$A = \nabla \boldsymbol{f} = \nabla \begin{pmatrix} u \\ v \end{pmatrix} = \begin{pmatrix} \nabla u \\ \nabla v \end{pmatrix} = \begin{pmatrix} \frac{\partial u}{\partial x} & \frac{\partial u}{\partial y} \\ \frac{\partial v}{\partial x} & \frac{\partial v}{\partial y} \end{pmatrix} = \begin{pmatrix} u_x & u_y \\ v_x & v_y \end{pmatrix}. \tag{2.60}$$

2.6 Computation of definite integrals*

Definition 2.5. Given a function $f(x)$, any differentiable function $F(x)$ is called a **primitive function** or an **anti-derivative** of $f(x)$ if

$$F'(x) = f(x), \tag{2.61}$$

or denoting the differential operator by D, $Df = f'$, then $DF = f$. I.e. "any differentiable function giving f when differentiated". Together with (A.1), we may write

$$D^{-1}Df(x) = f(x) + C, \quad DD^{-1}f(x) = f(x). \tag{2.62}$$

For computation of definite integrals in calculus, recourse is made to the following formula and one gets a feeling that one can always find a primitive function. But as we have seen, most primitive functions cannot be expressed in terms of elementary functions.

Corollary 2.4. (Computation of definite integrals) *Suppose $f : I \to \mathbb{R}$ is continuous on the open interval I and that for $\forall\,[a,b] \subset I$, F is a* **primitive function** *of f (on I) (i.e. a function such that $F'(x) = f(x)$). Then we may compute the value as*

$$\int_a^b f(x)\,\mathrm{d}x = F(b) - F(a). \tag{2.63}$$

We often denote the right-hand side by $[F(x)]_a^b$ or $F(x)|_a^b$.

Proof. By Corollary A.2, $\int_a^x f(t)\,\mathrm{d}t$ is a primitive function of $f(x)$, and so from the Newton-Leibniz rule (Corollary 1.5), it follows that

$$F(x) = \int_a^x f(t)\,\mathrm{d}t + c$$

(c being an integral constant). □

Exercise 2.33. For $m, n \in \mathbb{N}$ prove the following formulas, where δ_{mn} is the **Kronecker delta** (symbol) defined by

$$\delta_{mn} = \begin{cases} 1 & m = n, \\ 0 & m \neq n. \end{cases}$$

This already appeared in the statement of Theorem 1.17.

(i) $\displaystyle\frac{1}{2\pi}\int_{-\pi}^{\pi} \cos mt \cos nt\,\mathrm{d}t = \delta_{mn}$,

(ii) $\displaystyle\frac{1}{2\pi}\int_{-\pi}^{\pi} \sin mt \sin nt\,\mathrm{d}t = \delta_{mn}$,

(iii) $\displaystyle\int_{-\pi}^{\pi} \cos mt \sin nt\,\mathrm{d}t = 0$. This exercise gives a proof of Lemma 2.2 below.

Example 2.6. We evaluate the integral in Example 1.27 by an elementary method. Suppose $b, c \in \mathbb{R}$ satisfy $b^2 - 4c < 0$. Then we have

$$\int_{-\infty}^{\infty} \frac{1}{(x^2 + bx + c)^2}\,\mathrm{d}x = \frac{4\pi\sqrt{4c - b^2}}{(4c - b^2)^2}. \tag{2.64}$$

Proof. Recall the partial fraction expansion (1.229):

$$\frac{1}{(x^2+bx+c)^2} = \frac{-\frac{1}{4c-b^2}}{(x-\alpha)^2} + \frac{-\frac{2\sqrt{4c-b^2}}{(4c-b^2)^2}i}{x-\alpha} + \frac{-\frac{1}{4c-b^2}}{(x-\bar{\alpha})^2} + \frac{\frac{2\sqrt{4c-b^2}}{(4c-b^2)^2}i}{x-\bar{\alpha}}$$
$$= \frac{-\frac{1}{4c-b^2}}{(x-\alpha)^2} + \frac{-\frac{1}{4c-b^2}}{(x-\bar{\alpha})^2} - \frac{2\sqrt{4c-b^2}i}{(4c-b^2)^2}\frac{\alpha-\bar{\alpha}}{(x-\alpha)(x-\bar{\alpha})} \tag{2.65}$$

where $\alpha = \frac{-b+\sqrt{4c-b^2}i}{2}$ and $\alpha - \bar{\alpha} = \sqrt{4c-b^2}i$. Hence

$$\frac{1}{(x^2+bx+c)^2} = -\frac{1}{4c-b^2}\left(\frac{1}{(x-\alpha)^2} + \frac{1}{(x-\bar{\alpha})^2}\right) + \frac{2}{4c-b^2}\frac{1}{x^2+bx+c}. \tag{2.66}$$

Recalling

$$\int \frac{1}{x^2+bx+c}\,\mathrm{d}x = \int \frac{1}{\left(x+\frac{b}{2}\right)^2 + \sqrt{\frac{4c-b^2}{4}}^2}\,\mathrm{d}x$$
$$= \frac{2}{\sqrt{4c-b^2}}\arctan\frac{2x+b}{\sqrt{4c-b^2}} + C, \tag{2.67}$$

we integrate (2.66) to obtain

$$\int \frac{1}{(x^2+bx+c)^2}\,\mathrm{d}x = \frac{1}{4c-b^2}\left(\frac{1}{x-\alpha} + \frac{1}{x-\bar{\alpha}}\right)$$
$$+ \frac{4\sqrt{4c-b^2}}{(4c-b^2)^2}\arctan\frac{2x+b}{\sqrt{b^2-4c}} + C \tag{2.68}$$
$$= \frac{1}{4c-b^2}\frac{2x+b}{x^2+bx+c} + \frac{4\sqrt{4c-b^2}}{(4c-b^2)^2}\arctan\frac{2x+b}{\sqrt{b^2-4c}} + C.$$

Hence

$$\int_{-\infty}^{\infty} \frac{1}{(x^2+bx+c)^2}\,\mathrm{d}x \tag{2.69}$$
$$= \left[\frac{1}{4c-b^2}\frac{2x+b}{x^2+bx+c} + \frac{4\sqrt{4c-b^2}}{(4c-b^2)^2}\arctan\frac{2x+b}{\sqrt{b^2-4c}}\right]_{-\infty}^{\infty}$$
$$= \frac{4\sqrt{4c-b^2}}{(4c-b^2)^2}(\arctan(\infty) - \arctan(-\infty)),$$

which leads to (2.64).

2.6.1 *Line integrals and Green's formula**

Corollary 2.4 is the fundamental theorem in calculus giving a basic way of computing definite integrals as the difference of values of a primitive function at the end points. This may be thought of as a 1-dimensional integral amounting to the 0-dimensional one. In this way, it is generalized in the form of a 2-dimensional integral amounting to the 1-dimensional (line) integral, which is called Green's formula.

Theorem 2.12. (Green) *Let C be a piecewise smooth Jordan curve with D its interior. If functions $P(x,y), Q(x,y) \in C^1(\bar{D})$ and the integrals*

$$\int\int_D \frac{\partial P}{\partial x}\,\mathrm{d}x\mathrm{d}y, \quad \int\int_D \frac{\partial Q}{\partial y}\,\mathrm{d}x\mathrm{d}y$$

exist, then the line integral $\int_C P\,\mathrm{d}x + Q\,\mathrm{d}y$ along the positive direction of C exists and

$$\int\int_D \left(\frac{\partial P}{\partial x} - \frac{\partial Q}{\partial y}\right)\mathrm{d}x\mathrm{d}y = \int_C P\,\mathrm{d}y + Q\,\mathrm{d}x. \tag{2.70}$$

We prove this theorem only in the case of an ordinate set.

Definition 2.6. A planar domain D is called a **vertical ordinate set** if any vertical line (a line parallel to the y-axis) meets the boundary C of D at most two points. In this case, D can be expressed in the form $a \le x \le b,\ \phi_1(x) \le y \le \phi_2(x)$. Hereafter we assume that $\phi_1(x), \phi_2(x)$ are continuous.

Similarly, if any horizontal line (a line parallel to the x-axis) meets the boundary C of D at most two points, then D is called a **horizontal ordinate set**. In this case D can be expressed as $\psi_1(y) \le x \le \psi_2(y),\ c \le y \le d$ and we assume that $\psi_1(y), \psi_2(y)$ are continuous.

Proof of Theorem 2.12. Suppose $\bar{D}$ is vertically and horizontally ordinate set. Circumscribe $\bar{D}$ by a rectangle that meet at most one point and let vertical tangent points and horizontal tangent points be K, L and M, N, respectively. Then KML and KNL are expressed as one-valued functions $y = \phi_1(x)$ and $y = \phi_2(x)$ on $[a,b]$, respectively. Since $C = \text{arc } KML + \text{arc } LNK$, it follows that

$$\begin{aligned} \int_C Q\,\mathrm{d}x &= \int_{\text{arc } KML} Q\,\mathrm{d}x + \int_{\text{arc } LNK} Q\,\mathrm{d}x \\ &= \int_a^b Q(x,\phi_1(x))\,\mathrm{d}x - \int_a^b Q(x,\phi_2(x))\,\mathrm{d}x. \end{aligned} \tag{2.71}$$

By Theorem A.6,

$$\int\int_D -\frac{\partial Q}{\partial y}\,\mathrm{d}x\mathrm{d}y = -\int_a^b \int_{\phi_1(x)}^{\phi_2(x)} \frac{\partial Q}{\partial y}\,\mathrm{d}x\mathrm{d}y \tag{2.72}$$

$$= -\int_a^b [Q(x,y)]_{\phi_1(x)}^{\phi_2(x)}\,\mathrm{d}x = \int_a^b Q(x,\phi_1(x))\,\mathrm{d}x - \int_a^b Q(x,\phi_2(x))\,\mathrm{d}x,$$

which is $\int_C Q\,\mathrm{d}x$ by (2.71). Hence

$$\int\int_D \frac{\partial Q}{\partial y}\,\mathrm{d}x\mathrm{d}y = \int_C Q\,\mathrm{d}x. \tag{2.73}$$

Similarly, working with NKM, MLN, we obtain

$$\int\int_D \frac{\partial P}{\partial x}\,\mathrm{d}x\mathrm{d}y = \int_C P\,\mathrm{d}y. \tag{2.74}$$

Adding these completes the proof. □

2.7 Cauchy integral theorem

In Chapter I we emphasized the use of the Cauchy residue theorem and the reader may have got an impression that it is the fundamental theorem in complex analysis. Theoretically, however, the most fundamental principle is

Theorem 2.13. (Cauchy integral theorem) *If $f(z)$ is one-valued and analytic in a simply-connected domain $D \subset \mathbb{C}$, then for any piece-wise smooth closed Jordan curve $C \subset D$, we have*

$$\int_C f(z)\,\mathrm{d}z = 0. \tag{2.75}$$

Proof. Let $f = u + iv$. We apply Theorem 2.12 to

$$\int_C f(z)\,\mathrm{d}z = \int_C u\mathrm{d}x - v\mathrm{d}y + i\int_C v\mathrm{d}x + u\mathrm{d}y \tag{2.76}$$

to deduce ($X = \mathrm{Int}\, C \cup C$)

$$\int_C f(z)\,\mathrm{d}z = \int\int_X -\left(\frac{\partial v}{\partial x} + \frac{\partial u}{\partial y}\right)\mathrm{d}x\mathrm{d}y + i\int\int_X \left(\frac{\partial u}{\partial x} - \frac{\partial v}{\partial y}\right)\mathrm{d}x\mathrm{d}y, \tag{2.77}$$

in which both of the double integrals are 0 by the Cauchy-Riemann equation, completing the proof. □

To be able to apply Green's Theorem, we need to assume that u, v are of $C^1(X)$ class. There was tremendous endeavor made toward the generalization of the Cauchy integral theorem whose general form is the following.

Theorem 2.14. *Let C be any rectifiable Jordan closed curve with its finite interior D. If $f(z)$ is bounded and analytic on D and continuous on $\bar{D} = D \cup C$, then* (2.75) *holds true.*

The proof uses the method of approximating C by a polygon $C' \in D$ and proving that the integrals along C and C' can be made as close as we please in view of continuity of $f(z)$ on $\bar{D}$.

For practical use, however, integration paths are piecewise smooth ones combining polygons, arcs etc. and the functions considered are mostly smooth, we may take the use of the integral theorem for granted.

Theorem 2.13 is equivalent to the following two fundamental theorems corresponding to Corollary A.2 and Corollary 2.4, respectively. This indicates that the Cauchy integral theorem is of utmost importance in complex analysis but since primitive functions cannot in general be expressed in elementary functions, Theorem 2.16 is of little practical use for evaluating the value of an integral. It is residue calculus that shows its power in computing the values of integrals.

Theorem 2.15. (Existence of a primitive function) *If $f(z)$ is one-valued and analytic in a simply-connected domain $D \subset \mathbb{C}$, then there exists a function $F(z)$ analytic in D such that*

$$F'(z) = f(z). \tag{2.78}$$

A primitive function is unique up to a constant (integral constant).

As in calculus, the uniqueness is a consequence of the Newton-Leibniz rule.

Theorem 2.16. (Computation of definite integrals) *If $f(z)$ is one-valued and analytic in a simply-connected domain $D \subset \mathbb{C}$, then for any curve $C : z = z(t), t \in [a, b]$ in D of finite length, we have*

$$\int_C f(z)\,\mathrm{d}z = F(z(b)) - F(z(a)), \tag{2.79}$$

where $F(z)$ is a primitive function of $f(z)$.

The Cauchy integral theorem has a converse given by

Theorem 2.17. (Morera) *If $f(z)$ is continuous in a simply-connected domain $D \subset \mathbb{C}$ and*

$$\int_C f(z)\,\mathrm{d}z = 0 \tag{2.80}$$

for any piecewise smooth Jordan closed curve C in D, then $f(z)$ is analytic in D.

Proof. Fix a point z_0 and consider the indefinite integral

$$F(z) = \int_{z_0}^{z} f(w)\,\mathrm{d}w. \tag{2.81}$$

Since $F(z)$ does not depend on the choice of the path connecting z_0 and z (by the Cauchy integral theorem), it is uniquely defined. Since $f(z)$ is continuous on D, for any $z \in D$ and any $\varepsilon > 0$ there exists a $\delta > 0$ such that

$$|f(z+h) - f(z)| < \varepsilon \quad \text{for} \quad |h| < \delta.$$

By Theorem 2.16,

$$\frac{F(z+h) - F(z)}{h} = \frac{1}{h}\int_{z}^{z+h} f(w)\,\mathrm{d}w.$$

Substituting $f(z) = \frac{f(z)}{h}\int_z^{z+h} \mathrm{d}w$ from this, we deduce that

$$\left|\frac{F(z+h) - F(z)}{h} - f(z)\right| \le \frac{1}{|h|}\int_{z}^{z+h} |f(w) - f(z)|\,|\mathrm{d}w| < \frac{1}{|h|}\varepsilon|h| = \varepsilon. \tag{2.82}$$

Hence $F'(z) = f(z)$ and $F(z)$ is analytic in D. By Goursat's theorem (Theorem 2.1), $f(z)$ is also analytic in D. □

Remark 2.3. The above proof may be thought of as deducing Theorem 2.16 from Theorem 2.15 as well as giving a proof of Corollary A.2 by viewing all the variables as real.

Definition 2.7. A mapping f that preserve the angle including orientation is called a **conformal mapping**. More precisely, the angle formed by the tangents of two smooth curves meeting at z_0 is equal to the angle formed by the tangents of the curves mapped by f at the image point $w_0 = f(z_0)$.

Theorem 2.18. *If $f(z)$ is analytic in D, then at each $z_0 \in D$ for which $f'(z_0)$, $f(z)$ is conformal.*

Theorem 2.19. (Riemann mapping theorem) *Suppose $D \neq \mathbb{C}$ and $z_0 \in D$. Then there exists unique analytic function $f(z)$ that maps D one-to-one into the disc $|w| < 1$ such that $f(z_0) = 0$ and $f'(z_0) > 0$.*

Corollary 2.5. *An two simply-connected domains which are not $\mathbb{C}$ are mapped conformally on each other.*

Remark 2.4. The Riemann mapping theorem asserts the existence of a conformal mapping but does not give a way to construct it. In general constructing a conformal mapping is not easy but it is essential for applications. E.g. suppose $w = f(z)$ maps a domain D conformally in the upper half-plane $\mathcal{H} = \{w | \operatorname{Im} w > 0\}$ and that the complex potential $g(w)$ of a flow in $f(D)$ is known, then the complex potential of the flow in D is given by the composite function $(g \circ f)(z)$.

Example 2.7. Let

$$\mathrm{SL}_2(\mathbb{Z}) = \left\{ \gamma = \begin{pmatrix} a & b \\ c & d \end{pmatrix} \middle| a, b, c, d \in \mathbb{Z}, ad - bc = 1 \right\} \tag{2.83}$$

be the special linear group (or sometimes called the full modular group). The linear fractional transformation by an element $\gamma \in \mathrm{SL}_2(\mathbb{Z})$

$$\gamma(z) = \frac{az + b}{cz + d} \tag{2.84}$$

maps $\mathcal{H}$ conformally onto itself.

Proof. For the proof it suffices to verify

$$\operatorname{Im} \gamma(z) = \frac{\operatorname{Im} z}{|cz + d|^2} \tag{2.85}$$

and that

$$\frac{\mathrm{d}}{\mathrm{d}z} \gamma(z) = \frac{1}{(cz + d)^2}. \tag{2.86}$$

Theorem 2.20. (Multiply-connected domains) *Let C_0 be a piecewise smooth Jordan closed curve in which there are n Jordan closed curves $C_1, \ldots, C_n$ such that they do not intersect nor contain others. Let D be an $n+1$-multiply connected domain which is the interior of C_0 exclusive of the interiors of C_k's. All these curves are positively oriented.*

If $f(z)$ is one-valued and analytic in D, then

$$\int_{C_0} f(z)\,\mathrm{d}z = \sum_{k=1}^{n} \int_{C_k} f(z)\,\mathrm{d}z. \tag{2.87}$$

Corollary 2.6. (Contraction principle) *Let D be a domain encircled by a positively-oriented Jordan curve C. If $f(z)$ is one-valued and analytic in D except at the point z_0, then for any (positively oriented) closed curve γ in C encircling z_0 we have*

$$\int_{C_0} f(z)\,\mathrm{d}z = \int_{\gamma} f(z)\,\mathrm{d}z. \tag{2.88}$$

In particular we may take γ to be a small circle surrounding z_0.

This was referred to in § 1.2. If a function is expanded into a Laurent series, it can be integrated by this using (1.8).

Definition 2.8. The scalar product of gradients by itself ∇^2 is called the Laplace operator and denoted Δ:

$$\Delta = \frac{\partial^2}{\partial x^2} + \frac{\partial^2}{\partial y^2}. \tag{2.89}$$

Those functions are called harmonic functions which satisfy the Laplace equation

$$\Delta f = \frac{\partial^2 f}{\partial x^2} + \frac{\partial^2 f}{\partial y^2} = 0. \tag{2.90}$$

Harmonic functions are important for solving boundary value problems. It is often the case to find a harmonic function as the real or imaginary part of an analytic function according to the following

Theorem 2.21. (i) *If $f(z)$ is analytic in D, then both its real and imaginary parts are harmonic.*
(ii) *Suppose $u(z)$ is a (real) harmonic function in a simply-connected domain D. Let $C \subset D$ be any piecewise smooth Jordan arc connecting z and z_0. Then the function defined (up to an integral constant) by the integral*

$$v(z) = \int_C u_x\,\mathrm{d}y - u_y\,\mathrm{d}x \tag{2.91}$$

is harmonic in D and the function $f(z) = u(z) + iv(z)$ is analytic in D. $v(z)$ is called the **conjugate harmonic function** *of $u(z)$.*

Proof. (ii) The function $g(z) = u_x - iu_y$ is analytic in D. By Theorem 2.15, there exists an analytic primitive function $f(z) = \int g(z)\,\mathrm{d}z$. Since

$$f(z) = \int (u_x - iu_y)\,(\mathrm{d}x + i\mathrm{d}y) = \int u_x\,\mathrm{d}x + u_y\,\mathrm{d}y + i\int u_x\,\mathrm{d}y - u_y\,\mathrm{d}x$$
$$= \int \mathrm{d}u + i\int u_x\,\mathrm{d}y - u_y\,\mathrm{d}x = u(z) + iv(z) + c.$$

Hence by (i), $v(z) = \operatorname{Im} f(z)(+c)$ is harmonic in D. □

Exercise 2.34. Find the conjugate harmonic function in $\mathbb{C}$ of the following.
(i) $u(z) = x^2 - y^2$
(ii) $u(z) = \log(x^2 + y^2)$
(iii) $u(z) = \frac{x(x-1)+y^2}{(x-1)^2+y^2}$.

2.8 Cauchy integral formula

Subsequently as closed curves we consider only positively-oriented Jordan curves, referring to them as simple closed curves. The Cauchy integral formula (Theorem 1.9) with consequences is stated in §1.4.1, which we restate for the sake of completeness.

Theorem 2.22. (Cauchy integral formula) *Suppose $f(z)$ is analytic in the interior including boundary of a curve C. Then for any z in C*

$$f(z) = \frac{1}{2\pi i}\int_C \frac{f(w)}{w-z}\mathrm{d}w. \tag{2.92}$$

Proof. The integrand is analytic in D save for $w = z$, by the contraction principle Corollary 2.6, we obtain

$$\frac{1}{2\pi i}\int_C \frac{f(w)}{w-z}\mathrm{d}w = \frac{1}{2\pi i}\int_\gamma \frac{f(w)}{w-z}\mathrm{d}w$$

for the (positively-oriented) circle $\gamma : |w - z| = \rho$ in D. Since $f(w)$ is continuous in D, given $\varepsilon > 0$ we choose ρ small enough to ensure that $|f(w) - f(z)| < \varepsilon$ on γ. Hence

$$\begin{aligned}\frac{1}{2\pi i}\int_\gamma \frac{f(w)}{w-z}\mathrm{d}w &= \frac{1}{2\pi i}f(z)\int_\gamma \frac{1}{w-z}\mathrm{d}w + \frac{1}{2\pi i}\int_\gamma \frac{f(w)-f(z)}{w-z}\mathrm{d}w \\ &= f(z) + O\left(\varepsilon\rho\frac{1}{\rho}\right) = f(z) + O(\varepsilon),\end{aligned}$$

whence the assertion follows. □

Corollary 2.7. (Goursat) *If $f(z)$ is analytic in a domain D, then it has all orders of derivatives $f^{(k)}(z)$, which are also analytic in D, given by the Cauchy integral formula*

$$\begin{aligned}\frac{f^{(k)}(z)}{k!} &= \frac{1}{2\pi i}\int_C \frac{f(w)}{(w-z)^{k+1}}\mathrm{d}w \\ &= \frac{1}{2\pi i \cdot k!}\int_C \frac{\mathrm{d}^k}{\mathrm{d}z^k}\frac{f(w)}{w-z}\mathrm{d}w,\end{aligned}$$

where C is a closed Jordan curve contained in D.

Along with those consequences in §1.4.1, the following is often used in applications.

Corollary 2.8. (Maximum modulus principle) *If $f(z)$ is analytic and non-constant in a domain D, then $|f(z)|$ does not take its maximum inside D.*

Also if $f(z)$ is analytic in a bounded domain D and is continuous on the closed domain $\bar{D}$, then $|f(z)|$ takes its maximum on the boundary ∂D.

A generalization of the maximum modulus principle is the **convexity principle** known as the **Phragmén-Lindelöf theorem**. The most comprehensive account of the Phragmén-Lindelöf theorem can be found in [Rademacher (1973), Chapter 5, pp. 58-70] and also in [Titchmarsh (1939)]. Here we follow [Titchmarsh (1939), pp. 80-81] to prove the following simple version.

Theorem 2.23. *Suppose $f(s) = f(\sigma + it)$ is analytic and of order $O\left(e^{\varepsilon|t|}\right)$ for every $\varepsilon > 0$ in the strip $\sigma_1 \leq \sigma \leq \sigma_2$ and*

$$f(\sigma_1 + it) = O\left(|t|^{k_1}\right), \quad f(\sigma_2 + it) = O\left(|t|^{k_2}\right).$$

Then

$$f(\sigma + it) = O\left(|t|^{k(\sigma)}\right)$$

uniformly in $\sigma_1 \leq \sigma \leq \sigma_2$, $k(\sigma)$ being the linear function joining the points (σ_1, k_1), (σ_2, k_2).

Proof. First suppose that $k_1 = 0. k_2 = 0$, so that $f(s)$ is bounded on $\sigma = \sigma_1$, $\sigma = \sigma_2$. Let M be the upper bound of $f(s)$ on these lines and on the line segment $[\sigma_1, \sigma_2]$. Then for every $\varepsilon > 0$, the function $e^{\varepsilon is} f(s)$ is bounded in the half-strip $\sigma_1 \leq \sigma \leq \sigma_2$, $t > 0$.

For

$$\left|e^{\varepsilon is} f(s)\right| \leq e^{-\varepsilon t} |f(s)| \leq |f(s)| \leq M$$

on $\sigma = \sigma_1, \sigma_2$.

Since $\lim_{t \to \infty} e^{\varepsilon is} f(s) = 0$, we have

$$\left|e^{\varepsilon is} f(s)\right| \leq M \tag{2.93}$$

on $\sigma = \sigma_1, \sigma_2$, $t = T$ for $T > 0$ large enough. Hence, by the maximum modulus principle, (2.93) holds in the half-strip, whence

$$|f(s)| \leq M e^{\varepsilon t}.$$

Making $\varepsilon \to 0$, we conclude that $|f(s)| \leq M$ in the half-strip. Similarly, for $t < 0$, and we conclude the assertion in the general case.

□

2.9 Taylor expansions and extremal values

2.9.1 *Taylor expansions for real functions**

Theorem 2.24. (n-th Taylor expansion) *Suppose* $f \in C^n(V(x_0))$ *(n times continuously differentiable in a neighborhood of x_0). Then*

$$f(x) = \sum_{k=0}^{n-1} \frac{f^{(k)}(x_0)}{k!}(x - x_0)^k + R_n(x),$$

where the remainder term $R_n = R_n(x)$ is given by

$$R_n = R_n(x) = \frac{1}{(n-1)!}\int_{x_0}^{x} (x-t)^{n-1} f^{(n)}(t)\,\mathrm{d}t.$$

Proof. By integration by parts,

$$\begin{aligned} &R_n(x) \qquad (2.94)\\ &= \frac{1}{(n-1)!}\left[(x-t)^{n-1} f^{(n-1)}(t)\right]_{x_0}^{x} + \frac{1}{(n-2)!}\int_{x_0}^{x}(x-t)^{n-2} f^{(n-1)}(t)\,\mathrm{d}t\\ &= -\frac{f^{(n-1)}(x_0)}{(n-1)!}(x-x_0)^{n-1} + R_{n-1}(x). \end{aligned}$$

Hence inductively,

$$R_n(x) = -\sum_{k=1}^{n-1}\frac{f^{(k)}(x_0)}{k!}(x-x_0)^k + R_1(x).$$

Substituting

$$R_1(x) = \int_{x_0}^{x} f'(t)\,\mathrm{d}t = f(x) - f(x_0)$$

and moving the terms to the other side, the conclusion follows. □

Corollary 2.9. *If* $f(x) \in C^\infty$ *and*

$$\exists M > 0 \text{ s.t. } f^{(n)}(x) = O(M^n)$$

holds in an interval I, then $f(x)$ is **real analytic** *in that interval (can be expanded into a power series):*

$$f(x) = \sum_{n=0}^{\infty}\frac{f^{(n)}(x_0)}{n!}(x-x_0)^n,$$

which is called the **Taylor expansion** *around x_0. In particular, the Taylor expansion around $x_0 = 0$ is called the* **Maclaurin expansion**.

Proof. For any two points x_0, x on I, the remainder term of the n-th Taylor expansion is

$$\begin{aligned} R_n &= \left| \frac{1}{(n-1)!} \int_{x_0}^{x} (x-t)^{n-1} f^{(n)}(t)\, \mathrm{d}t \right| \\ &\leq \frac{1}{(n-1)!} \int_{x_0}^{x} \left| (x-t)^{n-1} f^{(n)}(t) \right| \mathrm{d}t \\ &= O\left(\frac{M^n}{(n-1)!} \int_{x_0}^{x} |x-t|^{n-1}\, \mathrm{d}t \right) = O\left(\frac{M^n}{n!} |x-x_0|^n \right). \end{aligned} \tag{2.95}$$

Hence, noting Exercise 2.35 and letting $n \to \infty$, we conclude the assertion. □

Exercise 2.35. Prove

$$\lim_{n \to \infty} \frac{z^n}{n!} = 0$$

for $\forall z \in \mathbb{C}$.

Exercise 2.36. Use the first mean value theorem for integrals to prove that the remainder term in the n-th Taylor expansion may be written in the form of **Roche-Schlömilch remainder term**: For $1 \leq \forall p \leq n$ (not necessarily an integer), with x_1 lying between x_0, x

$$R_n = \frac{1}{(n-1)!p} f^{(n)}(x_1)(x-x_0)^p (x-x_1)^{n-p}.$$

Exercise 2.37. Write down two special cases of the Roche-Schlömilch remainder term with $p = 1$ (the **Cauchy remainder term**) and with $p = n$ (the **Lagrange remainder term**).

Exercise 2.38. Prove that x_1 in the Roche-Schlömilch remainder term may be written as

$$x_1 = x_0 + \theta(x - x_0)$$

with $0 < \exists\theta < 1$. Further, putting $h = x - x_0$, then it can be also written as

$$R_n = \frac{1}{(n-1)!p} f^{(n)}(x_0 + \theta h)(1-\theta)^{n-p} h^n.$$

Work out with the Cauchy and Lagrange remainder terms.

- The expansion of e^x:

$$e^x = \sum_{n=0}^{\infty} \frac{x^n}{n!}.$$

For in view of $(e^x)^{(n)} = e^x$, it suffices to note that when x lies in any finite interval, $e^x = O(1)$ i.e. with $(M = 1)$.

- The expansions of $\sin x$, $\cos x$:

$$\sin x = \sum_{n=0}^{\infty} (-1)^n \frac{x^{2n+1}}{(2n+1)!}, \quad \cos x = \sum_{n=0}^{\infty} (-1)^n \frac{x^{2n}}{(2n)!}.$$

It suffices to note for $\forall x \in \mathbb{R}$ that $\left| \begin{pmatrix} \sin \\ \cos \end{pmatrix}^{(n)} (x) \right| = \left| \begin{pmatrix} \sin \\ \cos \end{pmatrix} (x + \frac{\pi}{2} n) \right| = O(1)$

Theorem 2.25. (second mean value theorem for integrals) *If $f, g(x) \in L^1$ and f is monotone on $I = [a, b]$, then*

$$\exists \xi \in I \ s.t. \int_a^b f(x) g(x) \mathrm{d}x = f(a) \int_a^\xi g(x) \mathrm{d}x + f(b) \int_\xi^b g(x) \mathrm{d}x.$$

Corollary 2.10.

$$g(x) \in C, \ f(x) \uparrow, \ f(x) \geq 0 \, on \, I = [a, b] \Rightarrow$$

$$\exists \xi \in I \ s.t. \int_a^b f(x) g(x) \mathrm{d}x = f(b) \int_\xi^b g(x) \mathrm{d}x.$$

Exercise 2.39. Prove Theorem 2.25 and its corollary. Use Theorem 2.25 to prove the expansion of e^x:

$$e^x = \sum_{n=0}^{\infty} \frac{x^n}{n!}.$$

2.9.2 *Taylor expansion*

Theorem 2.26. *If $f(z)$ is analytic at z_0, then $f(z)$ can be approximated to any degree of accuracy in the maximal circle $|z - z_0| = r$ of analyticity by the polynomial*

$$\sum_{k=0}^{n-1} \frac{f^{(k)}(z_0)}{k!} (z - z_0)^k \tag{2.96}$$

as follows.

$$f(z) = \sum_{k=0}^{n-1} \frac{f^{(k)}(z_0)}{k!}(z - z_0)^k + R_n(z),$$
$$R_n(z) = \frac{1}{2\pi i} \int_{|w-z_0|=\rho} \left(\frac{z - z_0}{w - z_0}\right)^n \frac{f(w)}{w - z_0}\, \mathrm{d}w, \tag{2.97}$$

where $0 < \rho < r$, $|z - z_0| < \rho$ *and*

$$|R_n(z)| < \left(\frac{|z - z_0|}{\rho}\right)^n \frac{\rho}{r - \rho} M(\rho), \quad M(\rho) = \max_{|w-z_0|=\rho} |f(w)|. \tag{2.98}$$

Theorem 2.27. (Cauchy-Taylor) *If* $f(z)$ *is analytic at* z_0, *then it can be expanded into the Taylor series in the maximal circle contained in the domain D of analyticity:*

$$f(z) = \sum_{n=0}^{\infty} a_n (z - z_0)^n,$$

where a_n *is given by (Corollary 2.7)*

$$a_n = \frac{f^{(n)}(z_0)}{n!} = \frac{1}{2\pi i} \int_C \frac{\mathrm{d}^n}{\mathrm{d}z_0^n} \frac{f(w)}{w - z_0}\, \mathrm{d}w,$$

C being any closed contour contained in D.

2.9.3 *Extremal values**

Theorem 2.28. (sign change of derivative) *Suppose* $f(x)$ *is continuous on* $[a, b]$ *and differentiable on* (a, b). *Then, if* $f'(x) > 0$ *on* (a, b), *then* $f(x)$ *is strictly monotone increasing, while if* $f'(x) < 0$ *on* (a, b), *then* $f(x)$ *is strictly monotone decreasing. Only the* **stationary points** *(i.e. the points for which* $f'(x) = 0$*) are the candidates for* **extremal points**. *Further, if* $f(x)$ *is of class* C^2, *and if* $f''(x_0) < 0$ *then* f *has a local maximum at* x_0, *and if* $f''(x_0) > 0$ *then* f *has a local minimum at* x_0.

Proof. The first half, i.e. if f' is of constant sign, then f is monotone, immediately follows from Theorem A.5. The second half geographically means that the slope of the tangent is parallel to the earth at the summits of mountains or at the bottoms of a valley. □

Exercise 2.40. Prove the following **Jordan's inequality**: For $0 \le x \le \frac{\pi}{2}$, we have

$$\sin x \ge \frac{2}{\pi} x \tag{2.99}$$

(equality holds if and only if and only if $x = 0,\ \frac{\pi}{2}$). Also prove that as $a \to \infty$,

$$\int_0^{\frac{\pi}{2}} e^{-a \sin\theta}\, d\theta \le \frac{\pi}{2a}\left(1 - e^{-a}\right) \to 0. \tag{2.100}$$

Solution. Let $f(x) = \sin x - \frac{2}{\pi}x$. Then $f'(x) = \cos x - \frac{2}{\pi}$. Letting θ be the unique number such that $\cos\theta = \frac{2}{\pi}$ or $\theta = \arccos\frac{2}{\pi}$, we see that f' changes the sigh from negative to positive at $x = \theta$ and so f is monotone increasing on $(0, \theta)$ and decreasing on $(\theta, \frac{\pi}{2})$. Since $f(0) = f\left(\frac{\pi}{2}\right) = 0$, the inequality follows. The integral is $\le \int_0^{\frac{\pi}{2}} e^{-\frac{2a}{\pi}\theta}\, d\theta$ by (2.99) and (2.100) follows.

Exercise 2.41. For $x > 0$ prove that

$$\frac{x}{x^2+1} < \arctan x < x.$$

Plot the graph of each member.

Exercise 2.42. Prove that $\log x < 1 - \frac{1}{x} < x - 1$, for $0 < x < 1$ and that $1 - \frac{1}{x} < \log x < x - 1$, for $x > 1$. Plot the graph of each member.

Exercise 2.43. For $|x| < \frac{\pi}{2}$ prove that

$$\sin x \le x \le \tan x$$

and plot the graphs of members of the inequalities.

Exercise 2.44. For $p > 1, \frac{1}{p} + \frac{1}{q} = 1$ prove that for $x \ge 0$

$$\frac{1}{p}x^p + \frac{1}{q} \ge x.$$

Then replace x by $xy^{-\frac{q}{p}}$ to deduce

$$\frac{1}{p}x^p + \frac{1}{q}y^q \ge xy \tag{2.101}$$

and plot the graphs of members of the inequalities.

Exercise 2.45. For n-dimensional (real) vectors $\boldsymbol{a} = \begin{pmatrix} a_1 \\ \vdots \\ a_n \end{pmatrix}$, $\boldsymbol{b} = \begin{pmatrix} b_1 \\ \vdots \\ b_n \end{pmatrix}$ define their p-norm and q-norm respectively by $X = |\boldsymbol{a}|_p = \left(\sum_{k=1}^n |a_k|^p\right)^{\frac{1}{p}}$, $Y = |\boldsymbol{b}|_q = \left(\sum_{k=1}^n |b_k|^q\right)^{\frac{1}{q}}$. Using (2.101) in Exercise 2.44, prove

$$\sum_{k=1}^n |a_k b_k| \le \frac{1}{p}|\boldsymbol{a}|_p^p + \frac{1}{q}|\boldsymbol{b}|_q^q. \tag{2.102}$$

Prove that (2.102) may be rewritten as

$$\sum_{k=1}^{n} |a_k b_k| \le \frac{1}{p} X^p \lambda^p + \frac{1}{q} Y^q \lambda^{-q}$$

for $\lambda > 0$. Prove that the function on the left $f(\lambda) := \frac{1}{p} X^p \lambda^p + \frac{1}{q} Y^q \lambda^{-q}$ has its minimum XY at $\lambda = \frac{Y^{\frac{1}{p}}}{X^{\frac{1}{q}}}$, thereby establishing **Hölder's inequality**

$$|\boldsymbol{a} \cdot \boldsymbol{b}| \le \sum_{k=1}^{n} |a_k b_k| \le |\boldsymbol{a}|_p |\boldsymbol{b}|_q. \tag{2.103}$$

Note that Hölder's inequality (2.103) reduces to the **Cauchy-Schwarz inequality** for $p = q = 2$.

Remark 2.5. Monotonicity test in Theorem 2.28 (also called the first derivative test) is to the effect that if $f(x)$ is of the form

$$f(x) = f(x_0) + (f'(x_0) + o(1))(x - x_0)$$

in a neighborhood of x_0 (mean value theorem), then ordering between $f(x)$ and $f(x_0)$ depends on the sign of $f'(x_0)$; if further, $f(x)$ satisfies

$$f(x) = f(x_0) + f'(x_0)(x - x_0) + \left(\frac{f''(x_0)}{2} + o(1)\right)(x - x_0)^2$$

(Taylor expansion) and $f'(x_0) = 0$, then the behavior of f at x_0 is determined by the sign of $f''(x_0)$ (the **second derivative test**).

This second derivative test also holds for functions in several variables: Since the Taylor expansion is of the form

$$f(\boldsymbol{x}) = f(\boldsymbol{x}_0) + \nabla f(\boldsymbol{x}_0)(\boldsymbol{x} - \boldsymbol{x}_0) + \frac{1}{2!}{}^t(\boldsymbol{x} - \boldsymbol{x}_0)(A + o(1))(\boldsymbol{x} - \boldsymbol{x}_0),$$

what corresponds to $f'(x_0) = 0$ is $\nabla f(\boldsymbol{x}_0) = \boldsymbol{o}$ or

$$\frac{\partial f}{\partial x_1}(\boldsymbol{x}_0) = \cdots = \frac{\partial f}{\partial x_n}(\boldsymbol{x}_0) = 0$$

and what corresponds to the sign of $f''(x_0)$ is the positive or negative definiteness of the **Hess matrix**

$$A = \left(\frac{\partial^2 f}{\partial x_i \partial x_j}(\boldsymbol{x}_0)\right).$$

The latter can be easily checked by checking the sign of n minors A_r that are formed from the blocks on the principal diagonal. E.g., for the function of two variables $f(x, y)$,

$$A = \begin{pmatrix} \frac{\partial^2 f}{\partial x^2}(\boldsymbol{x}_0) & \frac{\partial^2 f}{\partial x \partial y}(\boldsymbol{x}_0) \\ \frac{\partial^2 f}{\partial x \partial y}(\boldsymbol{x}_0) & \frac{\partial^2 f}{\partial y^2}(\boldsymbol{x}_0) \end{pmatrix} = \begin{pmatrix} a & b \\ b & c \end{pmatrix}.$$

Hence A is positive definite when the minors $A_r > 0$, i.e. a, $\begin{vmatrix} a & b \\ b & c \end{vmatrix}$ are both positive, in which case $f(\boldsymbol{x}) \geq f(\boldsymbol{x}_0)$, so that $f(\boldsymbol{x}_0)$ is a local minimum.

A is negative definite when $(-1)^r A_r > 0$, i.e. $a < 0$, $ac - b^2 > 0$ (i.e. discriminant $D := b^2 - ac < 0$), in which case $f(\boldsymbol{x}) \leq f(\boldsymbol{x}_0)$, so that $f(\boldsymbol{x}_0)$ is the local maximum.

Example 2.8. We find the extremal values of the function $f(x, y) = (y^2 - 1)\, e^{-x^2 - y^2}$. The stationary points are obtained from $\frac{\partial f}{\partial x}(\boldsymbol{x}_0) = \frac{\partial f}{\partial y}(\boldsymbol{x}_0) = 0$ and they are the three points $\begin{pmatrix} 0 \\ 0 \end{pmatrix}$, $\begin{pmatrix} 0 \\ \pm\sqrt{2} \end{pmatrix}$. At $\begin{pmatrix} 0 \\ \pm\sqrt{2} \end{pmatrix}$, we have $a = -\frac{2}{e^2} < 0$, $D = -\frac{16}{e^4} < 0$, where it has the local maxima.

Example 2.9. We find the extremal values of the function

$$f(x, y) = (x^2 + y^2)\, e^{x^2 - y^2}$$

by an elementary method of using the Taylor expansion of the exponential function only. We let $h, k \to 0$ independently. Then consider $f(x+h, y+k)$. Noting that

$$f(x + h, y + k) = (x^2 + y^2 + 2xh + 2yk + h^2 + k^2)\, e^{x^2 - y^2} e^{2xh - 2yk + h^2 - k^2},$$

we substitute the Taylor expansion of the exponential function, omitting those higher order infinitesimal terms,

$$
\begin{aligned}
& f(x+h, y+k)=e^{x^2-y^2}\left(x^2+y^2+2xh+2yk+h^2+k^2\right) \\
& \cdot\left(1+2xh-2yk+h^2-k^2+2x^2h^2-4xyhk+2y^2k^2+\cdots\right) \\
& =e^{x^2-y^2}\left(x^2+y^2+2xh+2yk+h^2+k^2\right) \\
& \cdot\left(1+2xh-2yk+(2x^2+1)h^2-4xyhk+(2y^2-1)k^2+\cdots\right) \\
& =e^{x^2-y^2}\left(x^2+y^2+2x(x^2+y^2+1)h-2y(x^2+y^2-1)k\right. \\
& +\left(4x^2+1+(x^2+y^2)(2x^2+1)\right)h^2-4(x^2+y^2)xyhk \\
& \left.+\left(-4y^2+1+(x^2+y^2)(2y^2-1)\right)k^2+\cdots\right) \\
& =f(\boldsymbol{x})+(\nabla f)\boldsymbol{h}+\frac{1}{2!}A[\boldsymbol{h}]+\cdots,
\end{aligned}
$$

say. Hence $\nabla f=(2x(x^2+y^2+1)e^{x^2-y^2}, -2y(x^2+y^2-1)e^{x^2-y^2})$ and the stationary points are $\begin{pmatrix}0\\0\end{pmatrix}, \begin{pmatrix}0\\\pm1\end{pmatrix}$. At $\begin{pmatrix}0\\\pm1\end{pmatrix}$, we have $\frac{1}{2!}A[\boldsymbol{h}]=2(h^2-k^2)e^{-1}$, so that they are saddle points. At $\begin{pmatrix}0\\0\end{pmatrix}$, we have $\frac{1}{2!}A[\boldsymbol{h}]=h^2+k^2>0$. Hence $\begin{pmatrix}0\\0\end{pmatrix}$ is the minimal point.

We give two important examples where finding extremal values play an essential role.

Example 2.10. We deduce the secular determinant for the molecular orbital Ψ consisting of n atomic orbitals:

$$
\Psi=\sum_{k=1}^{n} c_k\phi_k, \tag{2.104}
$$

where ϕ_k are atomic orbitals and c_k are (complex) coefficients. Let H denote the Hamiltonian of the molecule and let

$$
E=\frac{\int_{\mathbb{R}^n}\Psi H\Psi\,\mathrm{d}\tau}{\int_{\mathbb{R}^n}\Psi^2\,\mathrm{d}\tau}, \tag{2.105}
$$

where in general, Ψ is to be treated as a complex vector, in which case $\Psi H\Psi$ resp. Ψ^2 are to be regarded as $\bar{\Psi}H\Psi$ resp. $|\Psi|^2$ and the integrals are over $\mathbb{C}^n$. We write

$$
H_{ij}=H_{ji}=\int_{\mathbb{R}^n}\phi_i H\phi_j\,\mathrm{d}\tau, \quad S_{ij}=S_{ji}=\int_{\mathbb{R}^n}\phi_i\phi_j\,\mathrm{d}\tau. \tag{2.106}
$$

Then

$$E = \frac{\sum_{i,j=1}^{n} H_{ij} c_i c_j}{\sum_{i,j=1}^{n} S_{ij} c_i c_j} = \frac{H_{ii} c_i^2 + 2c_i \sum_{\substack{k=1 \\ k \neq i}}^{n} H_{ki} c_k + \cdots}{S_{ii} c_i^2 + 2c_i \sum_{\substack{k=1 \\ k \neq i}}^{n} S_{ki} c_k + \cdots} \tag{2.107}$$

for each i, $1 \leq i \leq n$. Applying the differentiation rule for the quotient in the form

$$\left(\frac{f}{g}\right)' = \frac{f'}{g} - \frac{f}{g}\frac{g'}{g},$$

we deduce that

$$\frac{\partial E}{\partial c_i} = \frac{2H_{ii} c_i + 2\sum_{\substack{k=1 \\ k \neq i}}^{n} H_{ki} c_k}{S_{ii} c_i^2 + 2c_i \sum_{\substack{k=1 \\ k \neq i}}^{n} S_{ki} c_k + \cdots} - E \frac{2S_{ii} c_i + 2\sum_{\substack{k=1 \\ k \neq i}}^{n} S_{ki} c_k}{S_{ii} c_i^2 + 2c_i \sum_{\substack{k=1 \\ k \neq i}}^{n} S_{ki} c_k + \cdots}, \tag{2.108}$$

whence

$$2H_{ii} c_i + 2\sum_{k \neq i}^{n} H_{ki} c_k - 2ES_{ii} c_i + 2E \sum_{k \neq i}^{n} S_{ki} c_k = 0,$$

i.e. the system of linear equations

$$\left(H_{ii} - S_{ii} E\right) c_i + \sum_{k \neq i}^{n} \left(H_{ki} - S_{ki} E\right) c_k = 0, \quad 1 \leq i \leq n. \tag{2.109}$$

For (2.109) to have a non-trivial solution c_i, the coefficient matrix must be singular, so that

$$\begin{vmatrix} H_{11} - S_{11}E & H_{12} - S_{12}E & \cdots & H_{1n} - S_{1n}E \\ H_{21} - S_{21}E & H_{22} - S_{22}E & \cdots & H_{2n} - S_{2n}E \\ \cdots & & & \\ H_{n1} - S_{n1}E & H_{n2} - S_{n2}E & \cdots & H_{nn} - S_{nn}E \end{vmatrix} = 0. \tag{2.110}$$

For a normalized molecule, we may suppose that $S_{ij} = \delta_{ij}$, (2.110) reduces to

$$\begin{vmatrix} H_{11} - E & H_{12} & \cdots & H_{1n} \\ H_{21} & H_{22} - E & \cdots & H_{1n} \\ \cdots & & & \\ H_{n1} & H_{n2} & \cdots & H_{nn} - E \end{vmatrix} = 0, \tag{2.111}$$

which is the secular determinant for Ψ.

Example 2.11. Given a probability distribution of an information system, $\{p_1, \ldots, p_n\}$, $0 < p_k < 1$,

$$\sum_{k=1}^{n} p_k = 1, \tag{2.112}$$

we find the values of p_k for which the **information entropy**

$$S = S(p_1, \ldots, p_n) = -\sum_{k=1}^{n} p_k \log p_k \tag{2.113}$$

attains its maximum. Here we apply the **principle of the entropy increase.** We apply the **Lagrange undetermined coefficient method.** Putting

$$L = L(p_1, \ldots, p_n, \lambda) = S(p_1, \ldots, p_n) + \lambda \left(\sum_{k=1}^{n} p_k - 1 \right), \tag{2.114}$$

where λ is a parameter, we may find the extremal points from the equation $\nabla L = \mathbf{o}$: Since

$$\begin{aligned} \nabla L &= \left(\frac{\partial L}{\partial p_1}, \ldots, \frac{\partial L}{\partial p_n}, \frac{\partial L}{\partial \lambda} \right) \\ &= \left(-\log p_1 - 1 + \lambda, \ldots, -\log p_n - 1 + \lambda, \sum_{k=1}^{n} p_k - 1 \right), \end{aligned} \tag{2.115}$$

it follows that $-\log p_k - 1 + \lambda = 0$, i.e. $\log p_k = \lambda - 1$, $p_k = e^{\lambda - 1}$. Substituting this in (2.112), we conclude that $\lambda - 1 = -\log n$, or $e^{\lambda-1} = \frac{1}{n}$, whence that

$$p_1 = \cdots = p_n = \frac{1}{n}. \tag{2.116}$$

Eq. (2.116) is in conformity with our intuition that the entropy becomes the maximum when all the variables have the same value. Consider, e.g. the casting of a dice.

2.10 Laurent expansions

Cf. §1.3.3.

Theorem 2.29 (Laurent expansion). *If $f(z)$ is (one-valued and) analytic in the annulus (ring-shaped domain) $D : r < |z - z_0| < R$ $(0 < r < R)$, then we have the Laurent expansion of $f(z)$:*

$$f(z) = \sum_{n=-\infty}^{\infty} a_n (z - z_0)^n, \tag{2.117}$$

where the n-th Laurent coefficient a_n is given by $(r < \rho < R)$

$$a_n = \frac{1}{2\pi i}\int_{|z-z_0|=\rho} \frac{f(z)}{(z-z_0)^{n+1}}\,\mathrm{d}z \tag{2.118}$$
$$= \frac{1}{2\pi i}\int_{|z-z_0|=\rho} \frac{1}{n!} f(z) \frac{\mathrm{d}^n}{\mathrm{d}z_0^n}(z-z_0)^{-1}\,\mathrm{d}z, \quad n \in \mathbb{Z}.$$

The Laurent series (2.117) converges uniformly in any annulus contained in D.

For $n \geq 0$, $a_n = \frac{1}{n!}f^{(n)}(z_0)$ are the Taylor coefficients. The negative power part $\sum_{n=-\infty}^{-1} a_n(z-z_0)^n$ is called the principal part of $f(z)$ at $z = z_0$.

In Definition 1.3, we dealt with the case where the principal part is finite but a proper perspective is given in this theorem, although in practice we use only the case of finite principal part.

Corollary 2.11. *If $f(z)$ is analytic on the unit circle $C : |z| = 1$, then we have the* **Fourier expansion** *of $f(e^{i\theta}) = u(\theta) + iv(\theta)$ as a complex series*

$$f(e^{i\theta}) = \sum_{n=-\infty}^{\infty} a_n e^{in\theta}, \tag{2.119}$$

where a_n is called the n-th Fourier coefficient given by

$$\frac{1}{2\pi}\int_0^{2\pi} f(e^{i\theta})e^{-in\theta}\,\mathrm{d}\theta = \frac{1}{2\pi}\int_0^{2\pi} (u(\theta) + iv(\theta))e^{-in\theta}\,\mathrm{d}\theta. \tag{2.120}$$

Proof. Since $f(z)$ is analytic in an annulus D containing the unit circle, it is expanded into a Laurent series

$$f(z) = \sum_{n=-\infty}^{\infty} a_n z^n. \tag{2.121}$$

Writing

$$z = e^{i\theta}, \tag{2.122}$$

(2.121) leads to (2.119). The n-th Laurent coefficient becomes (2.120) on substituting (2.122) into (2.118). □

Remark 2.6. Under the correspondence (2.122) we consider the composite function $\tilde{f}(z) = f(e^{2\pi iz})$ in $D \supset C$. On the unit circle, (2.119) may be viewed as a real Fourier series (2.204) for the real-valued periodic function $f(x)$ of period $2T = 1$. The Fourier series converges to $f(x)$ for a.a. x. Therefore we may view (2.119) as a complex extension of the Fourier expansion of f periodic of period 1.

Example 2.12. If $f(z)$ is analytic in $\bar{\mathcal{H}}$, where $\mathcal{H}$ is the upper half-plane in Remark 1.4 with bar indicating its closure, then we still want to have the Fourier expansion (2.119) for a periodic functions $\tilde{f}(z)$ of period 1; $\tilde{f}(z+1) = \tilde{f}(z)$. Similarly, in the case where $\tilde{f}(z)$ is analytic in $\mathcal{H}$ we still want to have (2.119). To justify this, note that under the correspondence (2.122)

$$q = e^{2\pi i z}, \quad z \in \mathcal{H}, \tag{2.123}$$

the upper half-plane is mapped into the inside the unit circle with the origin removed:

$$\mathcal{H} \leftrightarrow 0 < |q| < 1 \tag{2.124}$$

and the point ∞ at infinity corresponds to the origin. Cf. (1.23). We make a convention that we understand the behavior of $\tilde{f}(z)$ at ∞ through the behavior of $\tilde{f}(z)$ at 0.

We suppose $\tilde{f}(z)$ is meromorphic at the origin and consider the Laurent expansion of $\tilde{f}(z) = f(q)$ at $q = 0$, i.e. in an annulus $0 < |q| < 1$:

$$\tilde{f}(z) = f(q) = \sum_{n=-\infty}^{\infty} a_n q^n = \sum_{n=-\infty}^{\infty} a_n e^{2\pi i n z}, \tag{2.125}$$

where a_{-n} are 0 for some $M > 0$ onwards. We may speak of (2.125) as a Fourier expansion of $\tilde{f}(z)$ which is a substitute for the periodicity $\tilde{f}(z+1) = \tilde{f}(z)$.

Definition 2.9. If $f(z)$ is a meromorphic function in $\mathcal{H}$ having the Fourier expansion (2.125) and satisfying the relation

$$f\left(\frac{-1}{z}\right) = z^{2k} f(z), \tag{2.126}$$

then f is called a (holomorphic) modular form of weight $2k \in \mathbb{N}$ if $f(z)$ is analytic at ∞, i.e. the series (2.125) starts from $n = 0$. If $a_0 = 0$, then a modular form is called a cusp form.

Corollary 2.12. (L'Hospital's law) *Suppose $f(z)$ and $g(z)$ are analytic at z_0 and that $f(z)$ [resp. $g(z)$] has a zero of order m [resp. n] at z_0, respectively. Then*

$$\lim_{z \to z_0} \frac{f(z)}{g(z)} = \begin{cases} 0, & m > n \\ \dfrac{f^{(m)}(z_0)}{g^{(m)}(z_0)}, & m = n \\ \infty, & m < n. \end{cases} \tag{2.127}$$

2.11 Differential equations

In this section we develop the theory of differential equations **DEs** to such an extent that can be solved just by the inverse of the logarithmic differentiation, leaving higher degree DE in §1.11. First we recall the logarithm function.

2.11.1 *The logarithm function**

Definition 2.10. The logarithm function (the natural logarithm to the base e) is the inverse function to the exponential function:

$$y = e^x \iff x = \log y \quad (y > 0), \tag{2.128}$$

positivity being the consequence of the region of y.

Substituting the second into the first, we obtain

$$y = e^{\log y}.$$

Differentiating this by the chain rule whereby using $(e^x)' = e^x$, we have

$$1 = e^{\log y}(\log y)' = y(\log y)',$$

or $(\log y)' = \frac{1}{y}$. Changing x and y, we obtain the important formula

$$(\log x)' = \frac{1}{x}, \quad x > 0. \tag{2.129}$$

In case $x < 0$, we have $|x| = -x$ and by the chain rule again,

$$(\log |x|)' = (\log(-x))' = \frac{1}{-x}(-x)' = \frac{1}{x}.$$

Hence altogether,

$$(\log |x|)' = \frac{1}{x}, \quad x \neq 0 \tag{2.130}$$

or in terms of integrals,

$$\int \frac{1}{x}\,\mathrm{d}x = \log |x| + C, \quad x \neq 0, \tag{2.131}$$

where the integral constant C may be different according as $x > 0$ or $x < 0$.

Corollary 2.13. (Inverse operation of logarithmic differentiation) *Suppose* $f(x) \neq 0$. *Then*

$$(\log |f(x)|)' = \frac{f'}{f} \iff \int \frac{f'}{f}\,\mathrm{d}x = \log |f(x)| + C, \quad f(x) \neq 0, \tag{2.132}$$

2.11.2 *Autonomous DE*

The following proposition gives a basis for all exponential phenomena.

Proposition 2.1. *The solution of the DE* (**autonomous equation**) *in* $y = y(t)$

$$y' = \lambda y, \tag{2.133}$$

where λ *is a constant, is given by*

$$y(t) = e^{\lambda t} y(0). \tag{2.134}$$

Proof. The first proof uses the fact that (2.133) is of **variables separable type**. Supposing $y(t) \neq 0$, we obtain $\frac{y'}{y} = \lambda$. Integrating this by means of the inverse of the logarithmic differentiation, we obtain $\log|y| = \lambda t + C_1$. Exponentiating this, we have

$$y = Ce^{\lambda t}. \tag{2.135}$$

Putting $t = 0$, we see that $C = y(0)$. Hence (2.134) follows. The case $y = 0$ corresponds to $C = 0$.

The second proof turns out to be very useful not only for this but also for other types of DEs. We must predict the final form of the solution and dividing it by $e^{\lambda t}$. The derivative of $ye^{-\lambda t}$ is $e^{-\lambda t}(y' - \lambda y)$ whose second factor is 0 because of (2.133). Hence in view of the Newton-Leibniz rule, we conclude that $ye^{-\lambda t} = C$, which leads to (2.135).

The third proof uses the Laplace transform and is given in Exercise 1.17. □

2.11.3 *First-order reaction*

Definition 2.11. In an irreversible **first-order reaction** $A \to B$, where the reacting substance A changes into the resulting substance B, we denote by $[A] = [A](t)$, $[B] = [B](t)$ the molarity of A, B, respectively. Then the **rate of increase** $-\frac{d[A]}{dt}$ of B obeys the (first-order) **reaction equation**

$$-\frac{\mathrm{d}[A]}{\mathrm{d}t} = k[A], \tag{2.136}$$

where k is a positive constant called the **reacting constant**, which is determined by experiments. By Proposition 2.1, the solution is given by

$$[A] = [A](t) = e^{-kt}[A](0). \tag{2.137}$$

Example 2.13. In the case of a **radioactive decay**, we write $N = N(t)$ for the number of atoms in the radioactive substance in place of molarity in Definition 2.11 to rewrite (2.136) as

$$\frac{\mathrm{d}N}{\mathrm{d}t} = -kN, \tag{2.138}$$

where k is a positive constant called the **decay constant**. The solution is

$$N = N(t) = e^{-kt}N_0 = e^{-kt}N(0). \tag{2.139}$$

The period of time T needed for the number of atoms to become half of the initial number $N(0) = N_0$ is called a **half life** (half-value period). From $\frac{1}{2}N_0 = N_0e^{-kT}$ we deduce the important relation between the decay constant and the half life:

$$e^{-kT} = \frac{1}{2} \tag{2.140}$$

or

$$T = \frac{\log 2}{k} = \frac{0.693147\cdots}{k}. \tag{2.141}$$

Remark 2.7.

(i) Half-life means that the amount of the substance becomes half and there still remains half. In general, if we denote the period of time needed for the substance decreases to $1/a$ of the initial amount, then (2.142) reads

$$T_{1/a} = \frac{\log a}{k}. \tag{2.142}$$

Hence

$$T_{1/2a} = \frac{\log 2 + \log a}{k} = \frac{0.693147 + \log a}{k}. \tag{2.143}$$

In particular, $T_{1/4} = 2\frac{\log 2}{k} = 2T_{1/2}$, $T_{1/8} = 3\frac{\log 2}{k} = 3T_{1/2}$, meaning that the $1/4$th period is twice as much as the half life and the $1/8$th period is three times as much. E.g. in the case of cesium in Example 2.14, (ii), it will take 90 years for it to decrease to $1/8$th amount of the initial amount.

(ii) In literature one sometimes finds another formula called the rate of reaction formula, which follows by eliminating the reaction constant from (2.139) by the half-life. Expressing the reaction constant from (2.142),

$$k = \frac{\log 2}{T}. \tag{2.144}$$

Substituting this in (2.139), one obtains

$$N(t) = N_0e^{-\frac{t\log 2}{T}} = N(0)\left(\frac{1}{2}\right)^{t/T}. \tag{2.145}$$

Example 2.14.

(i) Iodine-131 decays into stable Xe-131: $^{131}\text{I} \to {}^{131}\text{Xe}$. The decay constant of ^{131}I is 0.00866(days). Hence

$$T = T_I = \frac{0.693147\cdots}{0.00866} = 80.04\cdots(\text{days}).$$

(ii) Cesium 137 (Cs-137) decays into semi-stable $Ba - 137m$: $^{137}\text{Cs} \to {}^{137m}\text{Ba}$, . The decay constant for Cs is $k = 2.31 \times 10^{-2}(\text{sec}^{-1})$. We find half-life of Cs, (2.142) amounting to

$$T = T_{Cs} = \frac{0.693147\cdots}{2.31} \times 10^2 = 30.00\cdots(\text{years}).$$

(iii) Strontium 90 (^{90}Sr) decays into yttrium ^{90}Y: $^{90}\text{Sr} \to {}^{90}\text{Y}$.
The decay constant for Sr-90 is $k = 7.85 \times 10^{-10}(\text{sec}^{-1})$. We find half-life of Sr-90, (2.142) amounting to

$$T = T_{Sr} = \frac{0.693147\cdots}{7.85} \times 10^{10} = 0.0882862\cdots \times 10^{10}(\text{sec}).$$

In years, this is 27.9740 $\cdots$, after dividing by 3153600.

(iv) In the case of plutonium 239 (^{239}Pu). The decay constant for Pu-239 is $k = 2.88 \times 10^{-5}(\text{year}^{-1})$. We find half-life of Pu, (2.142) amounting to

$$T = T_{Pu} = \frac{0.693147\cdots}{2.88} \times 10^5 = 24,100\ (\text{years}).$$

(v) Radon Rn = Rn-222 decays into Polonium Po = Po-218: $^{222}\text{Rn} \to {}^{218}\text{Po}$. The reaction continues and there appear four kinds of radioactive substances. The decay constant for Rn is $k = 1.81 \times 10^{-10}(\text{day}^{-1})$. We find half-life of Rn, (2.142) amounting to

$$T = T_{Rn} = \frac{0.693147\cdots}{1.81} \times 10 = 3.825\cdots(\text{day}).$$

Hence in 4 days, the amount becomes halved.

(vi) (2.144) may be used to compute the decay constant $k = k_{Yit}$ from half-life. E.g. consider the case of yttrium ^{90}Y as a continuation of Example 2.14. Yttrium−90 decays into zirconium−90; $^{90}\text{Y} \to {}^{90}\text{Zr}$. Its half-life is 64 hours. Hence the decay constant $k = \frac{0.693147\cdots}{64} = 1.08 \times 10^{-2}$ $(\text{hour})^{-1}$.

It is essential that one is not supposed to only remember such a relation but is supposed to be able to derive it from the original reaction equation by integration or by the Newton-Leibniz rule.

Carbon 14 is a radio-active isotope of the carbon atom which undergoes β decay and changes into Nitrogen 14 by releasing electrons and anti-electronic nutrino with radiating beta rays. The Carbon 14 is a threshold of unstable and stable carbon atoms and has been used extensively for age determination—the Carbon 14 method.

$$^{14}C \to {}^{14}N + e^- + \bar{\nu}_e. \tag{2.146}$$

Nitrogen 14 is a stable isotope of Nitrogen atom which occupies 99.63% of all isotopes.

The half-life of ^{14}C is known to be about 5730 years. Hence as in Example 2.14, (vi), we may compute the decay constant as follows. $k = \frac{0.693147\cdots}{5730} = 1.08 \times 10^{-2}\ (\text{year})^{-1}$.

2.11.4 *System of reactions*

Lemma 2.1. *Consider the system of differential equations (DE) with initial values $x(0)$, $y(0)$ given*

$$\begin{cases} \dfrac{dx}{dt} = -k_1 x \\ \dfrac{dy}{dt} = k_3 x - k_2 y, \end{cases} \tag{2.147}$$

where k_1, k_2, k_3 are constants. Their solutions are given by

$$x = e^{-k_1 t} x(0) \tag{2.134}$$

and

$$y = \frac{k_3 x(0)}{k_2 - k_1}\left(e^{-k_1 t} - e^{-k_2 t}\right) + y(0) e^{-k_2 t}, \quad k_1 \neq k_2 \tag{2.148}$$

while for $k_1 = k_2$

$$y = (k_3 x(0) t + y(0)) e^{-k_2 t}. \tag{2.149}$$

Proof. The first equation of (2.147) can be solved in three different ways (Proposition 2.1) to give (2.134).

The second becomes

$$\frac{dy}{dt} + k_2 y = k_3 x,$$

whence noting that

$$\frac{d}{dt}(ye^{k_2t}) = e^{k_2t}\left(\frac{dy}{dt} + k_2y\right) = e^{k_2t}k_3x = k_3x(0)e^{(k_2-k_1)t},$$

we obtain, for $k_1 \neq k_2$,

$$ye^{k_2t} = \frac{k_3x(0)}{k_2-k_1}e^{(k_2-k_1)t} + c_1,$$

$$y = y(t) = \frac{k_3x(0)}{k_2-k_1}e^{-k_1t} + c_1e^{-k_2t}.$$

Putting $t = 0$, we get

$$y(0) = \frac{k_3x(0)}{k_2-k_1} + c_1, \ c_1 = y(0) - \frac{k_3x(0)}{k_2-k_1}.$$

Hence (2.148).

If $k_1 = k_2$, we have $ye^{k_2t} = k_3x_0t + c_2$, $c_2 = y(0)$, and so (2.149) follows. □

Example 2.15. An irreversible first-order reaction $A \to B \to C$ in which the reacting substance A changes into the resulting substance C by way of the intermediate substance B obeys the system of DEs

$$\begin{cases} -\dfrac{\mathrm{d}[A]}{\mathrm{d}t} = k_1[A] \\ -\dfrac{\mathrm{d}[B]}{\mathrm{d}t} = k_2[B] - k_1[A]. \end{cases} \tag{2.150}$$

We solve this under the initial conditions $[A]\big|_{t=0} = [A](0)$, $[B]\big|_{t=0} = 0$ to obtain

$$[A] = [A](t) = e^{-k_1t}[A](0), \tag{2.151}$$

and for $k_1 \neq k_2$,

$$[B] = [B](t) = \frac{k_1}{k_2-k_1}(e^{-k_1t} - e^{-k_2t})[A](0), \tag{2.152}$$

while for $k_1 = k_2$, $[B](t) = [A](0)k_1te^{-k_1t}$.

Proof. This is Lemma 2.1 with $k_3 = k_1$. □

Example 2.16. We take up the successive reactions $^{90}Sr \to {}^{90}Y \to {}^{90}Zr$. in Example 2.14, (vi). The decay constant k_1 for ^{90}Y is $k_1 = 2.826 \times 10^{-6}$ $(\text{hour})^{-1}$. Substituting this and the value $k_2 = \frac{0.693147\cdots}{64} = 1.08 \times 10^{-2}$ $(\text{hour})^{-1}$ of the decay constant in (2.152) gives

$$N_2 = N_2(t) = \varkappa N_1(0) \tag{2.153}$$

where N_2 is the number of atoms of ^{90}Y while N_1 is that of ^{90}Sr and where

$$\varkappa = \frac{2.826 \times 10^{-6}}{1.08 \times 10^{-2} - 2.826 \times 10^{-6}} \times (e^{-2.826\times10^{-6}t} - e^{-1.08\times10^{-2}t}). \quad (2.154)$$

Pure ^{90}Sr contains almost no ^{90}Y, but in about a month, it comes to radio-active equilibrium and contains $\frac{1}{3900}$ ^{90}Y, which coincides with the theoretical value

$$\varkappa = 1/3820.65605096 \quad (2.155)$$

computed on the basis of (2.155).

Exercise 2.46. Solve the system of DEs in Lemma 2.1 using linear algebra.

2.11.5 *Other reactions*

Definition 2.12. As in Definition 2.11, we refer to an irreversible reaction $A \to B$ as a **$\varkappa$-th order reaction** , if the rate of increase $-\frac{d[A]}{dt}$ of B obeys the ($\varkappa$-th-order) **reaction equation**

$$-\frac{\mathrm{d}[A]}{\mathrm{d}t} = k[A]^{\varkappa}. \quad (2.156)$$

Example 2.17. In a second-order reaction suppose that it takes 10 minutes for the concentration of the reactant A to become half of the initial concentration. What will be the concentration after 30 minutes from the start of reaction? The second-order reaction is described by the reaction equation

$$-\frac{\mathrm{d}[A]}{\mathrm{d}t} = k[A]^2. \quad (2.157)$$

Solving this, we obtain

$$[A](t) = \frac{1}{kt + [A](0)^{-1}} = \frac{[A](0)}{[A](0)kt + 1}. \quad (2.158)$$

The condition states that

$$\frac{1}{2}[A](0) = [A](10) = \frac{1}{10k + [A](0)^{-1}},$$

whence

$$k = \frac{1}{10}[A](0)^{-1}.$$

Hence (2.158) assumes the form

$$[A](t) = \frac{[A](0)}{\frac{1}{10}t + 1}. \quad (2.159)$$

Hence $[A](30) = \frac{1}{4}[A](0)$. Also half-life T is found from (2.158) as follows.

$$\frac{1}{2}[A](0) = [A](T) = \frac{[A](0)}{[A](0)kT + 1},$$

whence

$$T = \frac{[A](0)}{k} \tag{2.160}$$

i.e. depending on the value of the initial molarity. Note the difference between this and the first-order reaction in Remark 2.7, (i).

Exercise 2.47. Solve the following reaction equations with initial concentration $[A]_0 = [A](0)$.

(i) The third order reaction

$$-\frac{\mathrm{d}[A]}{\mathrm{d}t} = k[A]^3. \tag{2.161}$$

(ii) The half order reaction

$$-\frac{\mathrm{d}[A]}{\mathrm{d}t} = k[A]^{1/2}. \tag{2.162}$$

2.11.6 *Poisson distribution*

Theorem 2.30. *The functions $x_k(t)$ satisfying the system of differential equations with initial values $x_0(0) = a > 0$, $x_k(0) = 0$, $k \in \mathbb{N}$*

$$\begin{cases} \dfrac{\mathrm{d}x_0}{\mathrm{d}t} = -\lambda x_0 \\ \dfrac{\mathrm{d}x_k}{\mathrm{d}t} = \lambda x_{k-1} - \lambda x_k, \end{cases} \tag{2.163}$$

is given by

$$x_k(t) = x_0(0)e^{-\lambda t}\frac{(\lambda t)^k}{k!}. \tag{2.164}$$

Proof. The system (2.163) with $k = 1$ is exactly (2.148) in Lemma 2.1 with $k_1 = k_2 = k_3 = \lambda$. Hence $x_0(t) = e^{-\lambda t}x_0(0)$ and

$$x_1(t) = x_0(0)e^{-\lambda t}(\lambda t). \tag{2.165}$$

From this inductively, we conclude (2.179).

□

Definition 2.13. The probability distribution with the density function

$$Po_k(t) = e^{-\lambda t}\frac{(\lambda t)^k}{k!} \tag{2.166}$$

is called the **Poisson distribution** $Po(k)$.

Example 2.18. In binary distribution with parameters n and p ($n \in \mathbb{N}, 0 < p < 1$ being small, $q = 1 - p$), if $n \to \infty$ under the condition that $np = m$, then we get the Poisson distribution:

$$P\{X = k\} = {}_nC_k p^k q^{n-k} \to Po(k) = e^{-m}\frac{m^k}{k!}, \quad k = 0, 1, \ldots. \tag{2.167}$$

This probability distribution appears when one assembles many events which rarely happen. Examples include the number of occurrences of earthquake, decay of radio-active substances, etc.

The Poisson distribution is the binary distribution with success probability small and number of trials large.

Exercise 2.48. Check that $Po(k)$ in (2.167) is a distribution function.

Solution. $\sum_{k=0}^{\infty} Po(k) = e^{-m}\sum_{k=0}^{\infty}\frac{m^k}{k!} = 1.$

2.11.7 *Decay of radio-active substances*

We denote by $P_n(t)$ the probability that n particles are released from a radio-active substance for the time period t on which we make the following assumptions. $P_0(t)$ is the probability that no particle is released in the period of length t.

(i) The numbers of released particles in non-overlapping time intervals are independent.

(ii) The probability that one particle is released from the substance in the time interval of length h, say $(t, t + h)$ is

$$P_1(h) = \lambda h + o(h), \tag{2.168}$$

where $o(h)$ is a function in h which is of smaller order than h as $h \to 0$.

(iii) The probability that more than one particle are released from the substance in the time interval $(t, t + h)$ is

$$\sum_{k=2}^{\infty} P_k(h) = o(h). \tag{2.169}$$

It follows from (ii) and (iii) that

$$\sum_{k=1}^{\infty} P_k(h) = \lambda h + o(h). \tag{2.170}$$

Since in the time period t, one of the decays $0, 1, 2, \ldots$ occur, it follows that

$$\sum_{k=0}^{\infty} P_k(t) = 1. \tag{2.171}$$

For $t = h$ small, combining (2.170) and (2.171) we conclude that

$$P_0(h) = 1 - \lambda h + o(h). \tag{2.172}$$

Next we deduce the formula for $P_n(t+h)$ the probability that n particles are released in the time interval $(0, t+h) = (0, t) \cup (t, t+h)$. There are $n+1$ mutually independent cases in which there are $n-k$ particles in the former time interval and k in the latter for $k = 0, \ldots, n$, it follows that

$$P_n(t+h) = \sum_{k=0}^{n} P_{n-k}(t) P_k(h) = P_n(t)P_0(h) + P_{n-1}(t)P_1(h) + \cdots. \tag{2.173}$$

Substituting (2.168) and (2.172) in (2.173) we conclude that

$$P_n(t+h) = P_n(t)(1 - \lambda h) + P_{n-1}(t)\lambda h + o(h) \tag{2.174}$$

for $n \in \mathbb{N}$. For $n = 0$, (2.174) should read

$$P_0(t+h) = P_0(t)(1 - \lambda h) + o(h). \tag{2.175}$$

Subtracting $P_n(t)$ ($n \geq 1$ and $n = 0$) from both sides of (2.174) and (2.175), we deduce that

$$P_n(t+h) - P_n(t) = -P_n(t)\lambda h + P_{n-1}(t)\lambda h + o(h) \tag{2.176}$$

and

$$P_0(t+h) - P_0(t) = -P_0(t)\lambda h + o(h). \tag{2.177}$$

This means that $P_n(t)$ satisfies the system of DE

$$\begin{cases} \dfrac{\mathrm{d}}{\mathrm{d}t} P_0(t) = -\lambda P_0(t) \\ \dfrac{\mathrm{d}}{\mathrm{d}t} P_n(t) = -\lambda P_n(t) + \lambda P_{n-1}(t). \end{cases} \tag{2.178}$$

(2.178) with $n = 1$ is the one in Lemma 2.1 with $k_1 = k_2 = k_3 = \lambda$ and so applies.

$$P_1(t) = \lambda t e^{-\lambda t}. \tag{2.179}$$

From this inductively, we get

$$P_n(t) = e^{-\lambda t}\frac{(\lambda t)^n}{n!}. \tag{2.180}$$

This is in conformity with (2.167) with $m = \lambda t$.

2.12 Inverse functions

2.12.1 *Inverse trigonometric functions**

Example 2.19. Simplify the expression $\cos(\arcsin x)$.

Solution. Recall that $\alpha = \arcsin x$ is equivalent to $x = \sin\alpha$ with $-\frac{\pi}{2} \le \alpha \le \frac{\pi}{2}$. What we want to simplify is $\cos\alpha$ which is ≥ 0 in $-\frac{\pi}{2} \le \alpha \le \frac{\pi}{2}$. Hence $\cos\alpha = \sqrt{1-\sin^2\alpha}$, and so $\cos(\arcsin x) = \sqrt{1-x^2}$.

Exercise 2.49. Prove the identity

$$\arcsin x + \arccos x = \frac{\pi}{2} \tag{2.181}$$

for $x \in [-1, 1]$. Note that for $0 < \alpha, \beta < \frac{\pi}{2}$, (2.181) amounts to the fact that for complementary angles (i.e. their sum is $\frac{\pi}{2}$) in a right-angled triangle, their sine and cosine values are equal.

Solution. Setting $\alpha = \arcsin x$, we see that it is equivalent to $x = \sin\alpha$ with $-\frac{\pi}{2} \le \alpha \le \frac{\pi}{2}$, where $\cos\alpha \ge 0$. Similarly, $\beta = \arccos x \iff x = \cos\beta$ with $0 \le \beta \le \pi$, where $\sin\beta \ge 0$. Hence $\cos\alpha = \sin\beta$.

By the above remark, we may contend that $\cos\alpha = \sqrt{1-x^2} = \sin\beta$. Hence by (1.25) and the Pythagoras theorem, $\sin(\alpha+\beta) = 1$. In the interval $-\frac{\pi}{2} \le \alpha+\beta \le \frac{3}{2}\pi$, there is a unique value $\frac{\pi}{2}$ of $\alpha+\beta$, whence we conclude (2.181).

Example 2.20. Prove the addition theorem for the tangent function

$$\tan(\alpha+\beta) = \frac{\tan\alpha + \tan\beta}{1 - \tan\alpha\tan\beta}. \tag{2.182}$$

Prove that

$$\arctan\frac{1}{x} + \arctan x = \frac{\pi}{2}. \tag{2.183}$$

Proof. Dividing the first formula in (1.25) by the second and then dividing both the numerator and the denominator by $\cos\alpha\cos\beta$, (2.182) follows.

In view of oddness of the tangent function, we may suppose $x > 0$ and in view of the reciprocals, we may further suppose that $0 < x \le 1$. The case $x = 1$ being clear, we suppose $0 < x < 1$. Putting $\alpha = \arctan x \iff x = \tan\alpha$, we have $0 < \alpha < \frac{\pi}{4}$ and $\beta = \arctan\frac{1}{x} \iff \frac{1}{x} = \tan\beta$, with $\frac{\pi}{4} < \beta < \frac{\pi}{2}$.

It then follows that $\tan(\alpha+\beta) = \infty$ with $\frac{\pi}{4} < \alpha + \beta < \frac{3\pi}{4}$, whence we conclude (2.183). □

Exercise 2.50. Prove the identity

$$\arctan\frac{1}{2} + \arctan\frac{1}{3} = \frac{\pi}{4}. \tag{2.184}$$

Solution. We set

$$\alpha = \arctan\frac{1}{2} \iff \frac{1}{2} = \tan\alpha, \quad 0 < \alpha < \frac{\pi}{4}$$
$$\beta = \arctan\frac{1}{3} \iff \frac{1}{3} = \tan\beta, \quad 0 < \beta < \frac{\pi}{4}.$$

By (2.182), we find that $\tan(\alpha+\beta) = \frac{\frac{1}{2}+\frac{1}{3}}{1-\frac{1}{2}\cdot\frac{1}{3}} = 1$.

Since

$$0 < \alpha + \beta < \frac{\pi}{2},$$

we may solve $\tan(\alpha+\beta) = 1$ to conclude (2.184).

Exercise 2.51. (i) Prove that $\arctan 2 + \arctan 3 = \frac{3}{4}\pi$.
(ii) Prove that $\arcsin\frac{12}{13} = \arccos\frac{5}{13}$.
(iii) Prove that $5\arctan\frac{1}{7} + 2\arctan\frac{3}{79} = \frac{\pi}{4}$.
(iv) Prove that $4\arctan\frac{1}{5} - \arctan\frac{1}{239} = \frac{\pi}{4}$
(v) Prove that $6\arctan\frac{1}{8} + 2\arctan\frac{1}{57} + \arctan\frac{1}{239} = \frac{\pi}{4}$
(vi) Prove that $4\arctan\frac{1}{5} - \arctan\frac{1}{70} + \arctan\frac{1}{99} = \frac{\pi}{4}$.

Exercise 2.52. Prove Euler's formula $\arctan\frac{1}{p} = \arctan\frac{1}{p+q} + \arctan\frac{q}{p^2+pq+1}$ for $p, q \in \mathbf{N}$.

Exercise 2.53. Prove that

$$\arctan x + \arctan y = \arctan\frac{x+y}{1-xy} \tag{2.185}$$

is equivalent to $xy \ne 1$ with the range of the argument to be checked. (2.183) may be thought of as the extremal case $xy = 1$.

2.12.2 *A. Gaudi and the catenary curve*

In this section we are going to find the equation for a catenary curve C, which is the form of a foldable string suspended at two edges. The inverse hyperbolic function appears which is a relative to inverse trigonometric functions. The defining equation of the catenary curve is

$$y = \cosh x = \frac{e^x + e^{-x}}{2}$$

cf. (1.64). It is remarkable that A. Gaudi used the catenary curve to plan his arches, which is the upside down form of the catenary. Let the lowest point of the curve be the origin O. For any point $P(x, y)$ on C, let $s = s(x)$ be the length of the arc OP. Then from § 1.13, we know that

$$\frac{\mathrm{d}s}{\mathrm{d}x} = \sqrt{1 + \left(\frac{\mathrm{d}y}{\mathrm{d}x}\right)^2}. \tag{2.186}$$

Let ϕ be the angle subtended by the tangent at P and the x-axis. Then

$$\frac{\mathrm{d}y}{\mathrm{d}x} = \tan\phi. \tag{2.187}$$

Now we are to express $\tan\phi$ in a different way. For this we consider the equilibrium of forces at P, where there are three forces exerting. The tension T exerting in the direction of the tangent and the horizontal tension H, which are in equilibrium with the weight G of the arc OP. Hence

$$H = T\cos\phi, \quad G = T\sin\phi \tag{2.188}$$

so that

$$\tan\phi = \frac{G}{H}. \tag{2.189}$$

Letting γ be the weight of the string per unit length, we have $G = \gamma s$. Hence

$$\tan\phi = \frac{s}{a}, \tag{2.190}$$

where $a = \frac{H}{\gamma}$. Substituting (2.190) in (2.187), we find that

$$\frac{\mathrm{d}y}{\mathrm{d}x} = \frac{s}{a}. \tag{2.191}$$

Substituting (2.191) in (2.186), we finally deduce that

$$\frac{\mathrm{d}s}{\mathrm{d}x} = \sqrt{1 + \left(\frac{s}{a}\right)^2} = \frac{\sqrt{s^2 + a^2}}{a}, \tag{2.192}$$

which is the variables separable type and can be integrated easily by writing

$$\frac{\mathrm{d}s}{\sqrt{s^2+a^2}} = \frac{\mathrm{d}x}{a}.$$

Integration gives

$$\sinh^{-1}\frac{s}{a} = \frac{x}{a} + C. \tag{2.193}$$

The integral constant $C = 0$. Hence

$$\sinh^{-1}\frac{s}{a} = \frac{x}{a}$$

or

$$\frac{s}{a} = \sinh\frac{x}{a} \tag{2.194}$$

Combining (2.191) and (2.194), we conclude that

$$\frac{\mathrm{d}y}{\mathrm{d}x} = \sinh\frac{x}{a}. \tag{2.195}$$

Integrating (2.195), we obtain

$$y = a\cosh\frac{x}{a} + C.$$

The integration constant can be found to be $-a$. Hence we finally obtain the equation for the catenary curve

$$y = a\cosh\frac{x}{a} - a. \tag{2.196}$$

Remark. (i) If in (2.196), we approximate the exponential function by its Taylor expansion, then

$$y = \frac{1}{2a}x^2 + O\left(x^4\right). \tag{2.197}$$

Hence if x is small enough, then the catenary can be approximated by the parabola. (ii) On p. 84 of [Ito (1968)], instead of the above DE (2.192), the author deduced

$$\frac{\mathrm{d}^2y}{\mathrm{d}x^2} = \frac{1}{a}\sqrt{1+\left(\frac{\mathrm{d}y}{\mathrm{d}x}\right)^2},$$

which looks very hard to integrate.

2.13 Rudiments of the Fourier transform

We assemble some basic facts on Fourier transforms at an elementary level, which gives rise to a heuristic proof of the Shannon sampling theorem.

In electrical engineering, with ω being the frequency, the variable is often denoted by $s = \sigma + j\omega$ (j being the imaginary unit) or $p = \sigma + j\omega$. The celebrated sampling theorem [Shannon (1949)] is essential in frequency analysis and in particular, in signal transmission. We state it here to motivate our analysis and we will come back to this as Theorem 2.33 below. For terminology cf. Definition 2.14.

Theorem 2.31. *If the function $f(t)$ is band-limited with band-length 2Ω:*

$$\hat{f}(\omega) = 0 \quad |\omega| \geq \Omega > 0, \tag{2.198}$$

then f is uniquely determined by its values (samples) at a sequence of equidistant points, by the Nyquist rate $\Delta t = \frac{1}{2\Omega}$ apart:

$$f(t) = \sum_{n=-\infty}^{\infty} f_n \mathrm{si}(2\pi\Omega(t - n\Delta t) = \sum_{n=-\infty}^{\infty} f(n\Delta t) \frac{\sin(2\pi\Omega(t - n\Delta t))}{2\pi\Omega(t - n\Delta t)}, \tag{2.199}$$

where

$$f_n = f(n\Delta t) = f\left(n\frac{1}{2\Omega}\right) \tag{2.200}$$

is the Nyquist sample.

As an example of the Nyquist rate, we refer to [Kumagai (2007), pp. 30-33], where the principle of CD is stated. The pitch of the sound can be divided into $2^{16} = 65536$ parts because a CD can record 16 convex-concave points as one information and transforms into 0 and 1 signal. The CD reads these 16 information 44100 times per second. The reason for this depends on the assumption that human ears can hear the sound whose frequencies are up to 20 kHz = 20000 Hz. Since the sound with frequency 20000 Hz, by definition, oscillates 20000 times per second and so more than this number of sampling is needed. And for stereo recording, we need twice as many, whence the sampling frequency 44100. Thus the sampling (Nyquist) rate $\frac{\pi}{T}$ is $1/44100$ and the input digital signal is restored by the sampling theorem above, [Takahashi (2011)], [Takahashi (2012)] etc.

The assumption that all the frequencies higher than 20000 Hz may be neglected is rather controversial. For recording simple conversation may not be needed more, but for music as supreme art, this omission can be a serious problem because what is missing is often more meaningful as art, cf. [Takahashi (2014)]. We recall that as soon as the natural real sound is transformed into digital signals, it is not the real sound but an approximation.

We use the argument from [Kanemitsu and Tsukada (2007), pp. 154-156] whose slight modification gives a proof of the above theorem.

Although a function $f(t)$ behaves differently on different parts of the t-domain (t-axis), with the aid of the Fourier transform $\mathcal{F}f(t) = \hat{f}(\omega)$ to be defined later, it may sometimes be expressed by a unique formula

$$f(t) = \frac{1}{2\pi} \int_{-\infty}^{\infty} \hat{f}(\omega)\, e^{i\omega t}\, \mathrm{d}\omega. \tag{2.201}$$

This is called the **Fourier Integral Theorem**. To be more precise, we have

Theorem 2.32. *If f, f' are piecewise continuous and*

$$\int_{-\infty}^{\infty} |f(t)|\, \mathrm{d}t < \infty,$$

then the Fourier integral theorem holds in the following form:

$$\frac{1}{2}\{f(t+0) + f(t-0)\} = \frac{1}{2\pi} \int_{-\infty}^{\infty} \hat{f}(\omega)\, e^{it\omega}\, \mathrm{d}\omega. \tag{2.202}$$

If we define the **inverse Fourier transform** $\mathcal{F}^{-1}g$ of g by

$$\left(\mathcal{F}^{-1}g\right)(\omega) = \frac{1}{2\pi} \int_{-\infty}^{\infty} g(t)\, e^{it\omega}\, \mathrm{d}t, \tag{2.203}$$

then we may express the Fourier Integral Theorem as $\mathcal{F}^{-1}\mathcal{F}f = f$ or $\hat{\hat{f}}(-t) = f(t)$.

(2.201) may be deduced formally from [Kanemitsu and Tsukada (2007), Theorem 7.2, p. 141] as follows and at the same time we will find a candidate for the Fourier transform $\hat{f}$.

Lemma 2.2. *The system of exponential functions $\{e^{2\pi int} | n \in \mathbb{Z}\}$ forms an ONS on $[0, 1]$.*

The proof is given in Exercise 2.33. For another ONS, cf. Theorem 1.17.

Hereafter, given a function f on $[-T, T]$, we consider its **periodic extension** $\bar{f}$ coinciding f on $[-T, T]$ and often denote it by the same symbol f: $\bar{f} = f$. It suffices to consider f over $|t| < T$. If f is piecewise smooth and continuous on $[-T, T]$, then its periodic extension f can be expanded into a **Fourier series**:

$$f(t) = \sum_{n=-\infty}^{\infty} c_n e^{i\lambda_n t}, \quad \lambda_n = \frac{2\pi}{2T}n, \tag{2.204}$$

where the nth **Fourier coefficient** c_n is given by

$$c_n = \frac{1}{2T}\int_{-T}^{T} f(t)\, e^{-i\lambda_n t}\, dt. \tag{2.205}$$

Indeed, if f is expanded in the form of (2.204), then the coefficient c_n must be equal to the integral $\frac{1}{2T}\int_{-T}^{T} f(t)\, e^{-i\lambda_n t}\, dt := a_n$, say, in (2.205). This follows from Lemma 2.2: By the change of variable $t = 2Tu$, we transform the integral into

$$a_n = \int_{-\frac{1}{2}}^{\frac{1}{2}} f(2Tu)\, e^{-2\pi i n u}\, du. \tag{2.206}$$

Substituting (2.204) and changing the order of integration and summation, we deduce that

$$a_n = \sum_{m=-\infty}^{\infty} c_m \int_{-\frac{1}{2}}^{\frac{1}{2}} e^{2\pi i (m-n) u}\, du \tag{2.207}$$

the right hand-side integral being 0 save for $m = n$. Hence $a_n = c_n$, i.e. (2.205) is valid.

Letting $T \to \infty$, $\quad n \to \infty$ in (2.205), we may contend that

$$2Tc_n \to \int_{-\infty}^{\infty} f(t)\, e^{-i\lambda_n t} dt =: \tilde{f}(\lambda_n), \tag{2.208}$$

say, and therefore

$$\begin{aligned} f(t) &= \frac{1}{2\pi}\sum_{n=-\infty}^{\infty} 2Tc_n e^{i\lambda_n t}(\lambda_{n+1} - \lambda_n) \\ &\to \frac{1}{2\pi}\sum_{n=-\infty}^{\infty} \tilde{f}(\lambda_n)\, e^{i\lambda_n t}\Delta\lambda_n \sim \frac{1}{2\pi}\int_{-\infty}^{\infty} \tilde{f}(\omega)\, e^{it\omega}\, d\omega, \end{aligned} \tag{2.209}$$

which is a heuristic proof of Theorem 2.32.

Viewing (2.209) as

$$f(t) = \frac{1}{2\pi}\int_{-\infty}^{\infty}\left(\int_{-\infty}^{\infty} f(t)\, e^{-it\omega}\, dt\right) e^{i\omega t}\, d\omega,$$

this may be thought of as giving the motivation for the definition of $\hat{f}(\omega)$, i.e. $\hat{f} = \tilde{f}$ in (2.208).

Also using the defining equation

$$\int_{-\infty}^{\infty} f(x)\,\delta(t-x)\,\mathrm{d}x = f(t)$$

for the delta function and one of its well known properties

$$\frac{1}{2\pi}\int_{-\infty}^{\infty} e^{ixt}\,\mathrm{d}x = \delta(t),$$

we can give the following simple proof.

$$\begin{aligned}
\frac{1}{2\pi}\int_{-\infty}^{\infty} (Ff)(y)\,e^{iyt}\,\mathrm{d}y &= \int_{-\infty}^{\infty} \frac{1}{2\pi}\left(\int_{-\infty}^{\infty} f(x)\,e^{-ixy}\,\mathrm{d}x\right) e^{iyt}\,\mathrm{d}y \\
&= \int_{-\infty}^{\infty} f(x)\,\mathrm{d}x \left(\frac{1}{2\pi}\int_{-\infty}^{\infty} e^{iy(t-x)}\,\mathrm{d}y\right) \\
&= \int_{-\infty}^{\infty} f(x)\,\delta(t-x)\,\mathrm{d}x = f(t).
\end{aligned}$$

This is a legitimate proof if the inversion of the order of integration is justified.

Thus we define

$$\hat{f}(\omega) = (\mathcal{F}f)(\omega) = \int_{-\infty}^{\infty} f(t)\,e^{-i\omega t}\,\mathrm{d}t, \tag{2.210}$$

to be paired with (2.203).

In view of the appearance of the factor $\frac{1}{2\pi}$ in the Fourier integral theorem, we often introduce normalization by distributing it to both transforms. The **symmetric pair** of the Fourier transform and the inverse Fourier transform is

$$\begin{aligned}
(\mathcal{F}f)(y) = \hat{f}(y) &= \frac{1}{\sqrt{2\pi}}\int_{-\infty}^{\infty} f(x)\,e^{-ixy}\mathrm{d}x, \qquad (2.211)\\
\left(\mathcal{F}^{-1}f\right)(y) &= \frac{1}{\sqrt{2\pi}}\int_{-\infty}^{\infty} f(x)\,e^{ixy}\mathrm{d}x.
\end{aligned}$$

Along with (2.211), we may also consider the case where there does not appear the preceding factor:

$$(\mathcal{F}f)(\omega) = \hat{f}(\omega) = \int_{-\infty}^{\infty} f(t)\,e^{-2\pi i\omega t}\mathrm{d}t, \tag{2.212}$$

$$\left(\mathcal{F}^{-1}\hat{f}\right)(t) = \int_{-\infty}^{\infty} \hat{f}(\omega)\,e^{2\pi it\omega}\mathrm{d}\omega. \tag{2.213}$$

Then Theorem 2.32 may be stated symbolically as

$$\mathcal{F}^{-1}\mathcal{F}f = f, \quad \mathcal{F}\mathcal{F}^{-1}f = f \quad \text{a.e.} \tag{2.214}$$

Fig. 2.1 Jean-Baptiste Josepf Fourier (1768-1830)

Proof of Theorem 2.31. Now we may give a proof of the sampling theorem. Correspondingly to (2.204) and (2.205), we have

$$\hat{f}(\omega) = \sum_{n=-\infty}^{\infty} c_n \, e^{2\pi i \lambda_n \omega}, \quad c_n = \frac{1}{2\Omega} \int_{-\Omega}^{\Omega} \hat{f}(\omega) \, e^{-2\pi i \lambda_n \omega} \, d\omega, \tag{2.215}$$

$$\lambda_n = \frac{1}{2\Omega} n,$$

where $\hat{f}$ is the Fourier transform of f. Then differently from (2.208), we have the equality

$$2\Omega c_n = \int_{-\Omega}^{\Omega} \hat{f}(\omega) \, e^{-2\pi i \lambda_n \omega} \, d\omega = \int_{-\infty}^{\infty} \hat{f}(\omega) \, e^{-2\pi i \lambda_n \omega} \, d\omega = f(\lambda_n). \tag{2.216}$$

Then correspondingly to (2.209), we deduce that

$$\hat{f}(\omega) = \sum_{n=-\infty}^{\infty} 2\Omega c_n e^{2\pi i \lambda_n \omega} \frac{1}{2\Omega} = \sum_{n=-\infty}^{\infty} f(\lambda_n) \, e^{2\pi i \lambda_n \omega} \frac{1}{2\Omega}. \tag{2.217}$$

Applying the Fourier inversion (2.213) bearing (2.198) in mind, we obtain

$$f(t) = \int_{-\infty}^{\infty} \hat{f}(\omega) \, e^{2\pi i t \omega} d\omega = \int_{-\Omega}^{\Omega} \hat{f}(\omega) \, e^{2\pi i t \omega} d\omega. \tag{2.218}$$

Substituting (2.217) into (2.218) and assuming that the *interchange of integration and summation is legitimate*, we obtain

$$f(t) = \sum_{n=-\infty}^{\infty} f(\lambda_n) \frac{1}{2\Omega} \int_{-\Omega}^{\Omega} e^{2\pi i (t+\lambda_n)\omega} d\omega. \tag{2.219}$$

Now (2.219) leads to (2.199), completing the proof. □

2.14 Paley-Wiener theorem and signal transmission

In this section we introduce the Paley-Wiener theorem (Theorem 1.27) and its application to signal transmission. First we prepare the relevant function. (1.238) reads

$$\frac{\sin z}{z} = \sum_{n=0}^{\infty} (-1)^n \frac{z^{2n}}{(2n+1)!}, \tag{2.220}$$

which is a generalization of the well-known fact

$$\lim_{z \to 0} \frac{\sin z}{z} = (\sin z)'|_0 = 1. \tag{2.221}$$

Defining the value at $z = 0$ of the function $\frac{\sin z}{z}$ to be 1 (i.e. removing the removable singularity), the new function, often denoted by $\mathrm{si}(z)$ (sinus cardinalis), is an entire function.

$$\mathrm{si}(z) = \sum_{n=0}^{\infty} (-1)^n \frac{z^{2n}}{(2n+1)!} = \begin{cases} \frac{\sin z}{z}, & z \neq 0 \\ 1, & z = 0. \end{cases} \tag{2.222}$$

The function $\mathrm{sinc}(t) = \mathrm{si}(\pi t)$ is introduced in (1.191).

Definition 2.14. Any $f \in L^p$ whose Fourier transform has a support (the domain where it is $\neq 0$) $[-T, T]$ is called **band-limited** with highest frequency T or of **band width** $2T$, i.e. the higher frequency is chopped off, which is often expressed as $2\pi\Omega$ or similarly. The sampling rate $\frac{2\pi}{2T} = \frac{1}{\Omega}$ is called the **Nyquist rate** and the sequence obtained using this rate is called **Nyquist samples**.

The sampling theorem (already stated as Theorem 2.31) is of utmost importance in signal transmission and in a sense an orthogonal expansion w.r.t. the system of sinc-functions (cf. Theorem 1.17). There are many references on it, cf. [Papoulis (1962), pp. 50-52], [Weaver (1983), pp. 117-120] etc. The following version is from [Papoulis (1962), pp. 50-52].

Theorem 2.33. (Sampling theorem) *If the signal $f(t)$ is band-limited, then f is given by*

$$f(t) = \sum_{n=-\infty}^{\infty} f_n \operatorname{sinc}\left(\frac{T}{\pi}\left(t - \frac{\pi}{T}n\right)\right) = \sum_{n=-\infty}^{\infty} f_n \frac{\sin(Tt - \pi n)}{Tt - \pi n}, \tag{2.223}$$

where

$$f_n = f\left(n\frac{\pi}{T}\right) \tag{2.224}$$

is the Nyquist sample, i.e., f is uniquely determined by its values (samples) at a sequence of equi-distant points, by the Nyquist rate $\frac{\pi}{T}$ apart.

Proof. We recall the Fourier expansion of $\hat{f}$ in the interval $(-T, T)$ in (2.215) in slight different notation, where f_n is $\frac{T}{\pi}c_n$ with c_n the n-th Fourier coefficient

$$\hat{f}(z) = \frac{\pi}{T} \sum_{n=-\infty}^{\infty} f_n e^{-in\frac{\pi}{T}z},$$

so that we may write

$$\hat{f}(z) = r_{2T}(z)\frac{\pi}{T} \sum_{n=-\infty}^{\infty} f_n e^{-in\frac{\pi}{T}z}$$

with the rectangular pulse function r_{2T}. Recalling the pair

$$r_{2T}(z)e^{-in\frac{\pi}{T}z} \leftrightarrow \frac{T}{\pi}\operatorname{si}(Tt - n\pi)$$

in [Chakraborty *et al.* (2009), (A.31), p. 217], we immediately conclude (2.223). □

2.14.1 *Fourier analysis again*

We make a review on modern theory of Fourier transforms based on the Lebesgue integration.

Let $L^p(D)$ denote the **(Banach) space of all p-th power integrable functions** (in the Lebesgue sense), where D indicates the real line $\mathbb{R}$ or an interval $I \subset \mathbb{R}$ with p-norm

$$||f||_p = \left(\frac{1}{\sqrt{2\pi}} \int_D |f(t)|^p \,\mathrm{dt}\right)^{1/p} \tag{2.225}$$

and $||f||_\infty = \lim_{p\to\infty} ||f||_p = \sup_{t\in\mathbb{R}} |f(t)|$. For any function $f \in L^1$, we define its **Fourier transform**

$$F(\omega) = \hat{f}(\omega) = \frac{1}{\sqrt{2\pi}} \int_{-\infty}^{\infty} f(t)e^{-i\omega t}\,\mathrm{d}t = \lim_{T\to\infty} \frac{1}{\sqrt{2\pi}} \int_{|t|\le T} f(t)e^{-i\omega t}\,\mathrm{d}t \tag{2.226}$$

for each frequency $\omega \in \mathbb{R}$. By this we go into the space of frequencies:

time domain $\to$ *frequency domain.*

If $f \in L^2$, the limit in (2.226) is to be understood in the L^2-norm sense.

The underlying principle that supports both approaches by the sampling and Bernstein polynomials is the Paley-Wiener theorem (= Theorem 1.27) which reads

Theorem 2.34. *An integral L^2 function $f = f(t)$ of exponential type $\le A$, $A > 0$ is of the form*

$$f(t) = \frac{1}{\sqrt{2\pi}} \int_{-A}^{A} F(\omega)e^{i\omega t}\,\mathrm{d}\omega, \tag{2.227}$$

where $F = F(\omega)$ is the Fourier transform (2.226) of f and lies in L^2.

The following theorem is fundamental in harmonic analysis and is a more general version of Theorem 2.32.

Theorem 2.35. (Carlson's theorem) *Let $f \in L^p(\mathbb{R})$ for some $p \in (1,\infty)$ have the Fourier transform $F(\omega) = \hat{f}(\omega)$. Then the* **Fourier inversion formula** *holds true:*

$$\frac{1}{\sqrt{2\pi}} \int_{-\infty}^{\infty} \hat{f}(\omega)e^{it\omega}\,\mathrm{d}\omega = \lim_{T\to\infty} \frac{1}{\sqrt{2\pi}} \int_{|\omega|\le T} F(\omega)e^{it\omega}\,\mathrm{d}\omega = f(t) \tag{2.228}$$

for **almost every** *(a.e.) $t \in \mathbb{R}$.*

I.e. *frequency domain* $\to$ *time domain.*

The discrete analogue of Theorem 2.35 reads

Theorem 2.36. (Discrete Carlson's theorem) *Let f be periodic of period 2π and $f \in L^p(\mathbb{R})$ for some $p \in (1,\infty)$ with Fourier coefficients $\hat{f}(n)$. Then the* **Fourier expansion** *holds true:*

$$\frac{1}{\sqrt{2\pi}} \sum_{n=-\infty}^{\infty} \hat{f}(n)e^{int} = \lim_{N\to\infty} \frac{1}{\sqrt{2\pi}} \sum_{|n|\le N} \hat{f}(n)e^{int} = f(t) \tag{2.229}$$

for **almost every** *(a.e.) $t \in \mathbb{R}$.*

We note that this *a.e.* (or often $a.a =$ almost all) notion is different from a daily-life interpretation of a.e. A.e. equality means equality that holds true except for a point set whose measure is 0, e.g. except for the set of all rationals, i.e. except for all numbers that are used daily. With this understanding, we come to know the limitation of the extrapolation etc.

Theorem 2.35 (2.228) then asserts that

$$f(t) = \frac{1}{\sqrt{2\pi}} \int_{-\pi\Omega}^{\pi\Omega} F(\omega) e^{it\omega} \, d\omega \tag{2.230}$$

holds true for a.a. values of t.

We may "prove" (2.230) by the following lemma.

Lemma 2.3. ([Hirano (2012), Lemma 4.2]) *The evaluation of the Dirichlet integral reads*

$$\lim_{t\to\infty} \int_{-\infty}^{\infty} \frac{\sin 2\pi t x}{\pi x} \varphi(x) \, dx = \varphi(0) \tag{2.231}$$

provided that φ is continuous at the origin.

Heuristic proof of (2.230). Substituting (2.226) for $F(\omega)$ on the left of (2.230), we obtain

$$\frac{1}{\sqrt{2\pi}} \int_{-\pi\Omega}^{\pi\Omega} F(\omega) e^{it\omega} \, d\omega = \frac{1}{2\pi} \int_{-\infty}^{\infty} f(\tau) \, d\tau \int_{-\pi\Omega}^{\pi\Omega} e^{i(t-\tau)\omega} \, d\omega \tag{2.232}$$

on changing the order of integration, which is not assured to hold, thus the proof is heuristic. The inner integral is $\frac{2\sin(t-\tau)\pi\Omega}{t-\tau}$. Hence by the change of variable $t - \tau = u$, we conclude that

$$\frac{1}{\sqrt{2\pi}} \int_{-\pi\Omega}^{\pi\Omega} F(\omega) e^{it\omega} \, d\omega = \frac{1}{2\pi} \int_{-\infty}^{\infty} f(t-u) \frac{2\sin \pi u \Omega}{u} \, du, \tag{2.233}$$

which is (2.230) in view of Lemma 2.3.

Note that [Splettstösser (1985), (2)] asserts the truth of (2.230) for **each** $t \in \mathbb{R}$ on account of Theorem 2.34. In view of this, Splettstösser contains the validity of (2.230) for all values in the definition of band-limitedness, which is a precursor of Theorem 2.33.

2.14.2 *Restoration of signals*

The appearance of the sinus cardinalis function si(t) in (2.231) suggests us a third heuristic proof of the sampling theorem.

Proof of Theorem 2.33. We view (2.223) as a discrete convolution

$$(f * \mathrm{si})(t) = \sum_{n=-\infty}^{\infty} f\left(\frac{\pi}{T}n\right) \mathrm{si}\left(T(t - \frac{\pi}{T}n)\right) = \sum_{n=-\infty}^{\infty} f\left(t - \frac{\pi}{T}n\right) \mathrm{si}\left(T\frac{\pi}{T}n\right). \tag{2.234}$$

Then noting that $\mathrm{si}(x) = 0$, $x = k\pi, k \neq 0$, $\mathrm{si}(0) = 1$, we conclude that the last expression in (2.234) is $= f(t)$.

Definition 2.15. Given two functions f, g, the integral (resp. the sum) of $f(t)g(x - t)$ over the whole domain of t is called their continuous (resp. discrete) **convolution** and denoted by $(f * g)(x)$. E.g. if $D = (a, b)$, we define

$$(f * g)(x) = \int_a^b f(t)g(x - t)\,\mathrm{d}t, \quad (f * g)(x) = \sum_{n=a}^{b} f(n)g(x - n). \tag{2.235}$$

If $D = (-\infty, \infty)$,

$$(f * g)(x) = \int_{-\infty}^{\infty} f(t)g(x - t)\,\mathrm{d}t, \quad (f * g)(x) = \sum_{n=-\infty}^{\infty} f(n)g(x - n) \tag{2.236}$$

if the limits exist. Cf. [Hirano (2012), Definition 5] for a continuous convolution.

Remark 2.8.

(i) In (2.234), use is made of the commutativity $f * g = g * f$, which is not *a priori* guaranteed, so that the proof is not rigorous as it stands.

(ii) The notation is slightly different. T in [Papoulis (1962), pp. 50-52], W in [Splettstösser (1985)] , and Ω in [Weaver (1983), pp. 117-120]:

[Papoulis (1962), pp. 50-52]	[Splettstösser (1985)]	[Weaver (1983), pp. 117-120]
T	W	Ω

$$T = \pi W = 2\pi\Omega.$$

In scientific disciplines, there are some empirical *threshold numbers* which are used often with great efficiency, although the reason why such thresholds appear is not theoretically clear.

The following theorem ([Weaver (1983), Theorem 5.2, pp. 119-120]) explains why the sampling rate may be taken as the **10 samples per cycle of the highest frequency**.

Theorem 2.37. (Real world sampling theorem) *If the function $f(t)$ is almost band-limited with band-length 2Ω and* **almost time-limited** *with time limit $2X$:*

$$\int_{-\infty}^{X} |f(t)|\, dt < \varepsilon, \quad \int_{X}^{\infty} |f(t)|\, dt < \varepsilon \tag{2.237}$$

then $f(t)$ can be recovered to any desired accuracy from its sampled values at uniformly-spaced intervals $|\Delta t| < \frac{1}{10\Omega}$ apart:

$$f(t) = \sum_{n=-M}^{M} f(n\Delta t) \frac{\sin(2\pi 5\Omega(t - n\Delta t))}{2\pi 5\Omega(t - n\Delta t)} + R_\varepsilon(t), \tag{2.238}$$

where

$$|R_\varepsilon(t)| << \varepsilon, \quad M\Delta t > X, \quad \Delta t < \frac{1}{10\Omega}. \tag{2.239}$$

Remark 2.9. An example of a function which is both almost band-limited and almost time-limited is the Gaussian function or its several-variables version

$$f(t) = e^{-at^2}, \quad a > 0. \tag{2.240}$$

This is a density function of the normal (Gaussian) distribution as well as an almost unique example of a rapidly decreasing function. It appears in our modular relation theory as the intrinsic function to the celebrated Riemann zeta-function.

Thus many research fields are interrelated in a hidden way through the Gaussian distribution.

Since the exponential function may be thought of as a degenerated form of the Bessel functions, one may think of a generalization of the presented ideas with the Bessel function instead.

Remark 2.10. Theorem 2.33 has an interesting history as is described in [Splettstösser (1985)]. We shall briefly state some highlights. It was stated in his 1949 paper [Shannon (1949)], which had been submitted in 1940, and he referred to it as **a result of common knowledge** and referred to Nyquist and Whittaker [Whittaker (1915)]. It is therefore sometimes called the Shannon-Whittaker theorem [Ruderman and Bialek (1999)]. In Whittaker the series (2.223) is known as *cardinal series*.

Appendix

A.1 Integration*

As we learned in calculus, if a function $f(x)$ is defined on an interval $I = [a, b]$, the area (with sign) of the figure formed by the graph of $y = f(x)$ and the lines $x = a$, $x = b$, $y = 0$ is called the **definite integral** with lower limit a, upper limit b (or the integral of f from a to b) and denoted by

$$\int_a^b f(x)\,\mathrm{d}x.$$

Mathematically, it is defined by the division of the interval.

Definition A.1. Let $I = [a, b]$ be a finite closed interval and let $f : I \to \mathbb{R}$ be bounded (i.e. $\exists M > 0\ \forall x \in I\,(|f(x)| < M)$). We consider a **division** Δ of I into m disjoint subintervals:

$$\Delta : a = x_0 < x_1 < \cdots < x_m = b.$$

We understand Δ means the division of I into the union of subintervals $I = \bigcup_{k=1}^{m} I_k, I_k = [x_{k-1}, x_k]$ as well as the set of all division points $\{x_0, \ldots, x_m\}$. Adding more division points gives rise to a refinement of I. We denote the maximum of the width (1-dimensional measure) of m subintervals $\mu(I_k) = x_k - x_{k-1}$ by $|\Delta|$ and call it the **size**, mesh or norm of Δ:

$$|\,\Delta\,| = \max\{\mu(I_k) \mid 1 \leq k \leq m\}.$$

From each I_k choose $\forall \xi_k \in I_k$ and let $\boldsymbol{\Xi} = (\xi_1, \ldots, \xi_m)$. Then form the finite sum

$$S(f) = S(\Delta, f) = S(\Delta, f, \boldsymbol{\Xi}) := \sum_{k=1}^{m} f(\xi_k)\,\mu(I_k)$$

and call it the **Riemann sum.**

If $S(\Delta, f, \Xi)$ approaches a value S say, independently of the choice of ξ_k as we make the size $|\Delta|$ of the division smaller (refining and making the size smaller), i.e.

$$\forall \varepsilon > 0, \ \exists \delta = \delta(\varepsilon) > 0 \ \forall \Delta \ \forall \Xi \ \left(|\Delta| < \delta \Rightarrow |S(\Delta, f, \Xi) - S| < \varepsilon \right),$$

we say that f is **integrable** (in the Riemann sense) on I, and call the value S the **definite integral** of f (with a and b lower and upper limit, respectively), denoted

$$\int_a^b f(x) \, \mathrm{d}x \quad \text{or} \quad \int_I f(x) \, \mathrm{d}x.$$

The value S may be interpreted to mean the area with sign of the figure formed by the graph of f and the x-axis (counting the area of that part which lies above the axis as plus and one under minus). In contrast to multiple integrals to be introduced presently, these are sometimes called single integrals.

Similarly, for a bounded function $z = f(x, y)$ in two variables defined on the 2-dimensional finite closed interval $\mathbb{I} \subset \mathbb{R}^2$ we make a division of $\mathbb{I}$ into subintervals to define the double integral of f

$$\int_{\mathbb{I}} f \, \mathrm{d}\boldsymbol{x} = \iint_{\mathbb{I}} f \, \mathrm{d}x \mathrm{d}y$$

which signifies the volume of the figure formed by the graphic surface $z = f(x, y)$ and the xy-plane.

We may also introduce the **triple integral**

$$\int_{\mathbb{I}} f \, \mathrm{d}\boldsymbol{x} = \iiint_{\mathbb{I}} f \, \mathrm{d}x \mathrm{d}y \mathrm{d}z$$

of a function $f(x, y, z)$ in three variables defined on a finite closed interval $\mathbb{I} \subset \mathbb{R}^3$ by making the division of the interval.

As can be seen from the definition by division into subintervals, the definite integral depends on the **local-global principle** to the effect that collecting local data $\mathrm{d}A = f(x) \, \mathrm{d}x$ gives rise to the totality $A = \int \mathrm{d}A = \int_a^b f(x) \, \mathrm{d}x$.

Subsequently, we make an intuitive discussion and as soon as we may express the data of infinitesimal part in the form

$$f(x) \, \mathrm{d}x$$

we immediately refer to the local-global principle to conclude the results. More rigorous treatment is possible and can be found in many textbooks.

In this book we assume the following theorem. What appears is the cases $n = 1, 2, 3$.

Theorem A.1. *Let $\mathbb{I} \subset \mathbb{R}^n$ be an n-dimensional finite closed interval and suppose $f : \mathbb{I} \to \mathbb{R}$ is continuous, then f is integrable on $\mathbb{I}$. In particular, on the special interval*

$$[a, b]^n = \left\{ \boldsymbol{x} = {}^t(x_1, \ldots, x_n) \mid a \leq x_k \leq b,\ 1 \leq k \leq n \right\}$$

f is integrable.

Corollary A.1. *Let $\mathbb{I} \subset \mathbb{R}^n$ be a finite closed interval and suppose f, g : $\mathbb{I} \to \mathbb{R}$ are continuous. Then*
(i) (*linearity*) *For c_1, $c_2 \in \mathbb{R}$,*

$$\int_{\mathbb{I}} (c_1 f + c_2 g)\, \mathrm{d}\boldsymbol{x} = c_1 \int_{\mathbb{I}} f\, \mathrm{d}\boldsymbol{x} + c_2 \int_{\mathbb{I}} g\, \mathrm{d}\boldsymbol{x}$$

(ii) (*monotonicity*) $f(\boldsymbol{x}) \leq g(\boldsymbol{x}) \quad (\forall \boldsymbol{x} \in \mathbb{I})$

$$\Rightarrow \int_{\mathbb{I}} f(\boldsymbol{x})\, \mathrm{d}\boldsymbol{x} \leq \int_{\mathbb{I}} g(\boldsymbol{x})\, \mathrm{d}\boldsymbol{x}$$

(iii) *The absolute value $|f| : \mathbb{I} \to \mathbb{R}_{\geq 0}$ of f is also integrable and*

$$\left| \int_{\mathbb{I}} f(\boldsymbol{x})\, \mathrm{d}\boldsymbol{x} \right| \leq \int_{\mathbb{I}} |f|(\boldsymbol{x})\, \mathrm{d}\boldsymbol{x}$$

where $|f|(\boldsymbol{x}) = |f(\boldsymbol{x})|$.
(iv) (*additivity with respect to intervals*) *If $\int_a^b f(x)\, \mathrm{d}x$ exists for $a < b$, we define*

$$\int_b^a f(x)\, \mathrm{d}x = -\int_a^b f(x)\, \mathrm{d}x.$$

Also we define

$$\int_a^a f(x)\, \mathrm{d}x = 0.$$

Then for a, b, $c \in \mathbb{R}$, we have

$$\int_a^b f(x)\, \mathrm{d}x + \int_b^c f(x)\, \mathrm{d}x + \int_c^a f(x)\, \mathrm{d}x = 0$$

in the sense that if two of the three integrals exist then the third exists and the equality holds,

Corollary A.2. (Fundamental theorem for infinitesimal calculus) *If* $f : I = [a, b] \to \mathbb{R}$ *is continuous, then putting*

$$F(x) = \int_a^x f(t)\,\mathrm{d}t$$

$(a \leq x \leq b)$, *F is of class C^1, and*

$$\frac{\mathrm{d}}{\mathrm{d}x} F(x) = \frac{\mathrm{d}}{\mathrm{d}x} \int_a^x f(t)\,\mathrm{d}t = f(x). \tag{A.1}$$

Exercise A.1. Prove Corollary A.2 in the same lines as in the proof of Theorem 2.17.

A.1.1 *Integration by parts and change of variables*

The definite integrals may be evaluated once we find a primitive function, but finding primitive functions almost all the time fails in the sense that they cannot be expressed in elementary functions (Liuoville's theorem). There are, however, two typical methods known for finding primitive functions.

Integration by parts

$$\int fg'\,\mathrm{d}x = fg - \int f'g\,\mathrm{d}x, \quad \int_a^b fg'\,\mathrm{d}x = [fg]_a^b - \int_a^b f'g\,\mathrm{d}x.$$

This follows from the formula for differentiation of products, Corollary A.2 and linearity.

Formula for change of variable

$$\int f(x)\,\mathrm{d}x = \int f(\phi(t))\,\phi'(t)\,\mathrm{d}t,$$

$$\int_a^b f(x)\,\mathrm{d}x = \int_\alpha^\beta f(\phi(t))\,\phi'(t)\,\mathrm{d}t, \; x = \phi(t),$$

where in the second formula, we suppose that ϕ has a continuous derivative monotone on $[\alpha, \beta]$ and f is continuous on $\phi([\alpha, \beta])$ and where $a = \phi(\alpha)$, $b = \phi(\beta)$.

Proof. The first formula follows from the differentiation of composite functions.

For the proof of the second formula, we note that on putting $F(y) = \int_a^y f(u)\,\mathrm{d}u$, $\Phi(t) = \int_\alpha^t f(\phi(u))\,\phi'(u)\,\mathrm{d}u$, we have from Corollary A.2,

$$F'(y) = f(y),\ \Phi'(t) = f(\phi(t))\,\phi'(t).$$

On the other hand, by differentiation of composite functions, it follows that

$$(F[\phi(t)])' = f(\phi(t))\,\phi'(t).$$

Hence it follows that

$$\Phi'(t) - (F[\phi(t)])' = 0.$$

By the Newton-Leibniz formula (p. 50), $\Phi(t) - (F[\phi(t)]) = \mathit{constant}$. Since $\Phi(\alpha) = 0$, $F[\phi(\alpha)] = F(a) = 0$, we conclude that this constant$= 0$, whence that $\Phi(t) = F[\phi(t)]$. In particular, putting $t = \beta$,

$$F(b) = F[\phi(\beta)] = \Phi(\beta).$$

□

We assemble some basic theorems on continuous functions.

Theorem A.2. (Weierstrass theorem) *A continuous function on a compact set X (typically on a bounded and closed interval $I = [a, b]$) has its maximum and minimum on X.*

Based on Theorem A.2, we have

Theorem A.3. (Intermediate value theorem) *If f is continuous on the closed interval $I = [a, b]$, then f takes on every value in* $[\min f(I), \max f(I)]$.

Theorem A.4. (The first mean value theorem for integrals) *Suppose $f(x)$, $g(x) \in C$ and that $g(x)$ is of constant sign (always positive or negative) on the interval $I = [a, b]$. Then*

$$\exists \xi \in I \text{ s.t. } \int_a^b f(x)\,g(x)\,\mathrm{d}x = f(\xi) \int_a^b g(x)\,\mathrm{d}x. \tag{A.2}$$

Proof. We may suppose that $g(x) > 0$. Let M, m be the maximum and minimum of $f(x)$ on I which exist by Theorem A.2. Hence we have $mg(x) \le f(x)g(x) \le Mg(x)$, $x \in I$, so that by monotonicity

$$m \int_a^b g(x)\,\mathrm{d}x \le \int_a^b f(x)\,g(x)\,\mathrm{d}x \le M \int_a^b g(x)\,\mathrm{d}x.$$

Hence putting $\lambda = \frac{\int_a^b f(x)g(x)\,\mathrm{d}x}{\int_a^b g(x)\,\mathrm{d}x}$, we have

$$\int_a^b f(x)\,g(x)\ \mathrm{d}x = \lambda \int_a^b g(x)\ \mathrm{d}x.$$

Since $\lambda \in [m, M] = f(I)$, it follows from Theorem A.3 that

$$\exists \xi \in I \text{ s.t. } f(\xi) = \lambda,$$

i.e. (A.2). □

Corollary A.3. *A special case of Theorem A.4 reads*

$$\int_a^b f(x)\mathrm{d}x = f(\xi)(b-a). \tag{A.3}$$

Theorem A.5. (Lagrange mean value theorem) *If f is continuous on $[a, b]$ and differentiable on (a, b), then there exists $\xi \in (a, b)$ such that*

$$f(b) - f(a) = f'(\xi)(b-a). \tag{A.4}$$

Figuratively, (A.4) *means that there exists a tangent line parallel to the chord joining two points $(a, f(a))$ and $(b, f(b))$.*

This can be proved by Exercise 2.38 etc. below.

Exercise A.2. Prove that (A.3) implies (A.1) and that (A.1) and Theorem A.5 imply (A.3).

Solution. Note that

$$\int_a^x f(t)\,\mathrm{d}t = f(\xi)(x-a)$$

by (A.3). Hence

$$\lim_{a\to x} \frac{1}{x-a}\int_a^x f(t)\,\mathrm{d}t = \lim_{a\to x} f(\xi) = f(x)$$

since f is continuous at $x \in [a, b]$, whence (A.1).

Conversely, (A.1) implies (2.63). By Theorem A.5, $F(b) - F(a) = F'(\xi)(b-a)$, which is $f(\xi)(b-a)$, by (A.1) and (A.3) follows.

A.1.2 *Multiple integrals**

The computation of single integrals can be reduced to finding a primitive function in view of Corollary 2.4 (which depends on the fundamental theorem of infinitesimal calculus). For n-ple integrals, we repeat the same process n-times. Such a computation is possible when the domain is the sum of some ordinate sets which are defined in Definition 2.6.

Theorem A.6. (double integral $\rightarrow$ repeated integral) *Suppose that the ordinate set $D: a \leq x \leq b, \quad \phi_1(x) \leq y \leq \phi_2(x)$ has the area and that f is continuous on D. Then the double integral $\iint f(x,y)\,\mathrm{d}x\mathrm{d}y$ may be computed as a repeated integral*

$$\iint_D f(x,y)\,\mathrm{d}x\mathrm{d}y = \int_a^b \left[\int_{\phi_1(x)}^{\phi_2(x)} f(x,y)\,\mathrm{d}y \right] \mathrm{d}x.$$

Remark A.1. In the above repeated integral, we view $f(x,y)$ as a function in y only and find the primitive function (in y). Then using Corollary 2.4 thereby substituting $\phi_1(x)$, $\phi_2(x)$, we find the function in x only and then find the primitive function and then use Corollary 2.4 again. The process is as "repeating anti-partial differentiation".

In case D is a horizontal ordinate set, we work with

$$\iint_D f(x,y)\,\mathrm{d}x\mathrm{d}y = \int_c^d \left[\int_{\psi_1(y)}^{\psi_2(y)} f(x,y)\,\mathrm{d}x \right] \mathrm{d}y.$$

Example A.1. Let D be the triangle with vertices at $(0,0)$, $(1,1)$, $(0,1)$. Then we evaluate the double integral $\iint_D e^{y^2}\,\mathrm{d}x\mathrm{d}y$. We may express D as a horizontal ordinate set $0 \leq x \leq y$, $0 \leq y \leq 1$ (if we use the vertical ordinate set $0 \leq x \leq 1$, $x \leq y \leq 1$, we cannot go on):

$$\iint_D e^{y^2}\,\mathrm{d}x\mathrm{d}y = \int_0^1 e^{y^2}\mathrm{d}y \int_0^y \mathrm{d}x = \int_0^1 y e^{y^2}\mathrm{d}y = \left[\frac{1}{2}e^{y^2}\right]_0^1 = \frac{1}{2}(e-1).$$

Example A.2. Let D be the square with vertices at $(1,0)$, $(0,1)$, $(-1,0)$, $(0,-1)$. Then we evaluate the double integral $\iint (x^2+y^2)\,\mathrm{d}x\mathrm{d}y$. $0 \leq x \leq$

$1,\ 0 \le y \le -x+1$:

$$\begin{aligned}
&\iint (x^2+y^2)\,\mathrm{d}x\mathrm{d}y \\
&= 4\int_0^1 \mathrm{d}x \int_0^{-x+1} (x^2+y^2)\,\mathrm{d}y = 4\int_0^1 \left[x^2 y + \frac{1}{3}y^3\right]_0^{-x+1} \mathrm{d}x \\
&= -4\int_0^1 \left((x-1+1)^2(x-1) + \frac{1}{3}(x-1)^3\right)\mathrm{d}x \\
&= -4\int_0^1 \left(\frac{4}{3}(x-1)^3 + 2(x-1)^2 + x - 1\right)\mathrm{d}x \\
&= -4\left[\frac{1}{3}(x-1)^4 + \frac{2}{3}(x-1)^3 + \frac{1}{2}(x-1)^2\right]_0^1 = \frac{2}{3}.
\end{aligned}$$

Theorem A.7. (change of variable for double integrals) *Suppose the vector-valued function* $\boldsymbol{\phi} : D \subset \mathbb{R}^2 \to \mathbb{R}^2,\ \boldsymbol{\phi}(u,v) = \begin{pmatrix} x(u,v) \\ y(u,v) \end{pmatrix}$ *satisfies the following conditions.* $\boldsymbol{\phi} \in C^1(D)$, *one-to-one on* D *and the Jacobian* $J_{\boldsymbol{\phi}} = \frac{\partial(x,y)}{\partial(u,v)} \neq 0$ *on* D. *Then by the change of variable*

$$\boldsymbol{x} = \boldsymbol{\phi}(\boldsymbol{t}),\ \boldsymbol{t} = \begin{pmatrix} u \\ v \end{pmatrix} \in D,$$

we have

$$\begin{aligned}
\iint_{\boldsymbol{\phi}(D)} f(\boldsymbol{x})\,\mathrm{d}\boldsymbol{x} &= \int_D f(\boldsymbol{\phi}(\boldsymbol{t}))\,|J_{\boldsymbol{\phi}}|\,\mathrm{d}\boldsymbol{t} \\
&= \iint_D f(x(u,v), y(u,v)) \left|\frac{\partial(x,y)}{\partial(u,v)}\right| \mathrm{d}u\mathrm{d}v,
\end{aligned}$$

where D *is the domain on the uv-plane corresponding to the domain* D *on the xy-plane. The formula also holds true if* $J_{\boldsymbol{\phi}} = 0$ *on subsets with measure* 0 *of* D.

Lemma A.1. *Suppose the functions* $x = x(u,v),\ y = y(u,v)$ *are of class* C^1 *on a domain* D. *Then we have*

$$\mathrm{d}x\mathrm{d}y = \frac{\partial(x,y)}{\partial(u,v)}\,\mathrm{d}u\mathrm{d}v, \tag{A.5}$$

where the computation of differential forms is to follow the rules $\mathrm{d}v\mathrm{d}u = -\mathrm{d}u\mathrm{d}v$ *and* $\mathrm{d}u\mathrm{d}u = \mathrm{d}v\mathrm{d}v = 0$, *where the first shows a relationship between areas with sign of two infinitesimal parallelograms with* $\mathrm{d}x$, $\mathrm{d}y$ *as adjacent sides, while the second means the areas of degenerated parallelograms.*

Explanation. By the results of § 2.5,

$$dx = \frac{\partial x}{\partial u}\,du + \frac{\partial x}{\partial v}\,dv,\ dy = \frac{\partial y}{\partial u}du + \frac{\partial y}{\partial v}dv.$$

Hence

$$\begin{aligned} dxdy &= \frac{\partial x}{\partial u}\frac{\partial y}{\partial v}dudv + \frac{\partial y}{\partial u}\frac{\partial x}{\partial v}dvdu = \left(\frac{\partial x}{\partial u}\frac{\partial y}{\partial v} - \frac{\partial y}{\partial u}\frac{\partial x}{\partial v}\right)dudv \\ &= \frac{\partial\,(x,y)}{\partial\,(u,v)}dudv. \end{aligned}$$

Now Theorem A.7 follows from Lemma A.1 in view of the local-global principle.

Example A.3. Let D denote the unit disc (with radius 1 and center at the origin). Then we evaluate the integral

$$\iint_{\phi(D)} e^{x^2+y^2}\,dxdy.$$

By the polar coordinate $\begin{pmatrix} x \\ y \end{pmatrix} = \boldsymbol{z} = \phi(r,\theta) = \begin{pmatrix} r\cos\theta \\ r\sin\theta \end{pmatrix}$

$$\phi(D) \leftrightarrow D = \left\{ \begin{pmatrix} r \\ \theta \end{pmatrix} \middle|\, 0 \le r \le 1,\ 0 \le \theta \le \pi \right\}.$$

Since the Jacobian is $\frac{\partial(x,y)}{\partial(r,\theta)} = \begin{vmatrix} \cos\theta & -r\sin\theta \\ \sin\theta & r\cos\theta \end{vmatrix} = r$ it vanishes only at the origin. Substituting $dxdy = rdrd\theta$,

$$\begin{aligned} \iint e^{x^2+y^2}\,dxdy = \iint e^{r^2}\,rdrd\theta &= \int_0^{2\pi} d\theta \int_0^1 re^{r^2}dr = \int_0^{2\pi}\left[\frac{1}{2}e^{r^2}\right]_0^1 d\theta \\ &= \pi\,(e-1). \end{aligned}$$

Exercise A.3. Let D be the domain on the xy-plane with the boundaries $x = 1,\ y = x,\ y = -x^2$.
(i) Then find the area of D. Also find the domain $\phi^{-1}(D)$ on the uv-plane which maps onto D under the transformation $\begin{pmatrix} x \\ y \end{pmatrix} = \boldsymbol{z} = \phi(u,v) = \begin{pmatrix} u \\ -u^2+v \end{pmatrix}$ whence find the area of D again.
(ii) Evaluate $\iint \frac{dx\,dy}{(x-y+1)^2}$.

Exercise A.4. For the domain D with boundaries $x = 0,\ y = 0,\ x + y = 1$, evaluate

$$\iint \exp\left(\frac{x-y}{x+y}\right)dxdy.$$

Exercise A.5. Evaluate the integral in Exercise A.4 by the change of variable $\begin{pmatrix} x \\ y \end{pmatrix} = z = \phi(u,v) = \begin{pmatrix} \frac{1}{2}(u+v) \\ \frac{1}{2}(u-v) \end{pmatrix}$.

Exercise A.6. Prove the formula

$$\Gamma(\alpha)\,\Gamma(\beta) = \Gamma(\alpha+\beta)\,\mathrm{B}(\alpha,\beta), \tag{A.6}$$

whence in particular

$$\Gamma\left(\frac{1}{2}\right) = \sqrt{\pi}, \tag{A.7}$$

or the value of the probability integral $\int_0^\infty e^{-x^2}\,\mathrm{d}x = \frac{\sqrt{\pi}}{2}$. Cf. Example 1.13.

Solution. The gamma function, being the Mellin transform of e^{-t} mentioned in Example 2.5. (ii), is defined by the **Eulerian integral of the second kind**

$$\Gamma(s) = \int_0^\infty e^{-t} t^{s-1}\,\mathrm{d}t \tag{A.8}$$

for $\Re s = \sigma > 0$. This improper integral is absolutely and uniformly convergent in the wide sense in $\sigma > 0$, whence it follows that $\Gamma(s)$ is analytic in the right half-plane $\sigma > 0$. Let the **beta function** $\mathrm{B}(\alpha,\beta)$ be defined by the **Eulerian integral of the first kind**

$$\mathrm{B}(\alpha,\beta) = \int_0^1 t^{\alpha-1}(1-t)^{\beta-1}\,\mathrm{d}t, \quad \Re\alpha > 0, \Re\beta > 0. \tag{A.9}$$

First, in (A.9), put $t = \sin^2\theta$ to obtain

$$\mathrm{B}(\alpha,\beta) = 2\int_0^{\frac{\pi}{2}} \sin^{2\alpha-1}\theta\,\cos^{2\beta-1}\theta\,\mathrm{d}\theta. \tag{A.10}$$

If in (A.8), we put $t = x^2$, then

$$\Gamma(s) = 2\int_0^\infty x^{2s-1} e^{-x^2}\,\mathrm{d}x,$$

whence for $\Re\alpha > 0$, $\Re\beta > 0$,

$$\begin{aligned}
\Gamma(\alpha)\,\Gamma(\beta) &= 4\int_0^\infty x^{2\alpha-1}e^{-x^2}\,\mathrm{d}x \int_0^\infty y^{2\beta-1}e^{-y^2}\,\mathrm{d}y \\
&= 4\lim_{X\to\infty}\left(\int_0^X x^{2\alpha-1}e^{-x^2}\,\mathrm{d}x \int_0^X y^{2\beta-1}e^{-y^2}\,\mathrm{d}y\right) \\
&= 4\lim_{X\to\infty}\int_0^X\int_0^X x^{2\alpha-1}y^{2\beta-1}\,e^{-(x^2+y^2)}\,\mathrm{d}x\,\mathrm{d}y \\
&= 4\lim_{X\to\infty}\iint_D x^{2\alpha-1}y^{2\beta-1}\,e^{-(x^2+y^2)}\,\mathrm{d}x\,\mathrm{d}y, \qquad \text{(A.11)}
\end{aligned}$$

where $D = \left\{ \binom{x}{y} \middle| 0 \le \sqrt{x^2+y^2} \le X \right\}$. By the change of variable $x = r\cos\theta,\ y = r\sin\theta$, we have the correspondence

$$D \leftrightarrow \tilde{D} = \left\{ \begin{pmatrix} r \\ \theta \end{pmatrix} \middle| 0 \le r \le X,\ 0 \le \theta \le \frac{\pi}{2} \right\}.$$

where the absolute value of the Jacobian of this transformation is $\left| \frac{\partial(x,y)}{\partial(r,\theta)} \right| = r$. Hence

$$\begin{aligned}
\Gamma(\alpha)\,\Gamma(\beta) &= 4 \lim_{X\to\infty} \iint_{\tilde{D}} r^{2\alpha+2\beta-2}\, e^{-r^2} \sin^{2\alpha-1}\theta\, \cos^{2\beta-1}\theta\, r\, \mathrm{d}r\mathrm{d}\theta \\
&= 2\int_0^\infty r^{2\alpha+2\beta-1} e^{-r^2}\, \mathrm{d}r \cdot 2\int_0^{\frac{\pi}{2}} \sin^{2\alpha-1}\theta\, \cos^{2\beta-1}\theta\, \mathrm{d}\theta \\
&= \Gamma(\alpha+\beta)\, \mathrm{B}(\alpha,\beta)
\end{aligned}$$

by (A.10) above. This completes the proof of (A.6).

Putting now $\alpha = \beta = \frac{1}{2}$, we obtain

$$\Gamma\left(\frac{1}{2}\right)^2 = \Gamma(1)\, 2\int_0^{\frac{\pi}{2}} \mathrm{d}\theta = \pi,$$

whence (A.7) follows. It follows that

$$\int_{-\infty}^{\infty} e^{-\frac{x^2}{2}}\, \mathrm{d}x = \sqrt{2\pi},$$

which is used to normalize the distribution function of the Gaussian (or normal) distribution.

Remark A.2. An ordinary procedure for proving (A.7) is to use (A.11) for $\alpha = \beta = \frac{1}{2}$

$$\left(\int_0^\infty e^{-x^2}\, \mathrm{d}x \right)^2 = 4 \lim_{R\to\infty} \left(\int_0^R r\, e^{-r^2}\, \mathrm{d}r \int_0^{\frac{\pi}{2}} \mathrm{d}\theta \right) = 4 \left[-\frac{1}{2} e^{-r^2} \right]_0^\infty \frac{\pi}{2} = \pi.$$

Thus we see that if we generalize the problem by introducing the parameters α and β, we get a wider perspective.

A.2 Answers and hints

Exercise 1.2 A more common procedure is the use of the recurrence formula for $n \ge 2$:

$$I_n := \int \sin^n x\, \mathrm{d}x = \frac{n-1}{n} I_{n-2} - \frac{1}{n} \sin^{n-1} x \cos x + C. \tag{A.12}$$

This may be proved by writing

$$I_n = I_{n-2} - \int \cos x \sin^{n-2} x \cos x \, dx.$$

The second integral may be transformed by integration by parts into

$$\frac{1}{n-1} \sin^{n-1} x \cos x + \frac{1}{n-1} I_n.$$

Substituting this in the previous equality, we deduce (A.12).

To deduce Wallis' formula, we use (A.12) in the form

$$J_n := \int_0^{\frac{\pi}{2}} \sin^n x \, dx = \frac{n-1}{n} J_{n-2} = \frac{n-1}{n} \frac{n-3}{n-2} J_{n-4}. \tag{A.13}$$

Noting that

$$\int_0^{\frac{\pi}{2}} \sin x \, dx = 1,$$

$$\int_0^{\frac{\pi}{2}} \sin^2 x \, dx = \int_0^{\frac{\pi}{2}} \frac{1-\cos 2x}{2} \, dx = \left[\frac{2x - \sin 2x}{4} \right]_0^{\frac{\pi}{2}} = \frac{1}{2} \frac{\pi}{2},$$

we obtain **Wallis' formula**

$$\int_0^{\frac{\pi}{2}} \sin^n x \, dx = \frac{n-1}{n} \frac{n-3}{n-2} \cdots \frac{1}{2} \frac{\pi}{2} \tag{A.14}$$

for n even and

$$\int_0^{\frac{\pi}{2}} \sin^n x \, dx = \frac{n-1}{n} \frac{n-3}{n-2} \cdots \frac{2}{3} \tag{A.15}$$

for n odd.

We note the equality (for $a \geq 2$)

$$\int_0^{\frac{\pi}{a}} \sin^n x \, dx = \int_0^{\frac{\pi}{2}} \cos^n x \, dx - \int_0^{\frac{\pi}{2} - \frac{\pi}{a}} \cos^n x \, dx \tag{A.16}$$

in which the sine may be replaced by the cosine. Hence the following evaluation is essential in this exercise:

$$\int_0^{\frac{\pi}{3}} \sin^n x \, dx, \int_0^{\frac{\pi}{6}} \sin^n x \, dx, \int_0^{\frac{\pi}{4}} \sin^n x \, dx.$$

By the formula in (A.14) it is easy to see that

$$\int_0^{\frac{\pi}{6}} \sin^4 x \, dx = \frac{1}{8} \left[\frac{1}{4} \sin 4x - 2 \sin 2x + 3x \right]_0^{\frac{\pi}{6}} = \frac{1}{8} \left(-\frac{7}{8} \sqrt{3} + \frac{\pi}{2} \right) \tag{A.17}$$

and

$$\int_0^{\frac{\pi}{3}} \cos^4 x \, dx = \frac{1}{8}\left[\frac{1}{4}\sin 4x + 2\sin 2x + 3x\right]_0^{\frac{\pi}{3}} = \frac{1}{8}\left(\frac{7}{8}\sqrt{3} + \pi\right). \quad \text{(A.18)}$$

These are consistent with (A.16) with $a = 6$. These may be checked by (A.12):

$$\int_0^{\frac{\pi}{6}} \sin^4 x \, dx = \frac{3}{4}\int_0^{\frac{\pi}{6}} \sin^2 x \, dx - \frac{1}{4}[\sin^3 x \cos x]_0^{\frac{\pi}{6}} = \frac{3}{4}\frac{1}{4}\left(\frac{\pi}{3} - \frac{\sqrt{3}}{2}\right) - \frac{1}{4}\frac{1}{8}\frac{\sqrt{3}}{2}.$$

Similarly,

$$\int_0^{\frac{\pi}{6}} \sin^6 x \, dx = -\frac{1}{32}\left[\frac{1}{6}\sin 6x - \frac{3}{2}\sin 4x + \frac{15}{2}\sin 2x - 10x\right]_0^{\frac{\pi}{6}} \quad \text{(A.19)}$$
$$= \frac{1}{32}\left(-3\sqrt{3} + \frac{5\pi}{3}\right).$$

This may be checked by (A.12):

$$\int_0^{\frac{\pi}{6}} \sin^6 x \, dx = \frac{5}{6}\int_0^{\frac{\pi}{6}} \sin^4 x \, dx - \frac{1}{6}[\sin^5 x \cos x]_0^{\frac{\pi}{6}}$$
$$= \frac{5}{6}\frac{1}{8}\left(-\frac{7}{8}\sqrt{3} + \frac{\pi}{2}\right) - \frac{1}{6}\left(\frac{1}{2}\right)^5 \frac{\sqrt{3}}{2}.$$

(B) (i) $\frac{1}{32}\left(-3\sqrt{3} + \frac{5}{3}\pi\right)$ (ii) $\frac{1}{32}\left(\frac{9}{2}\sqrt{3} + \frac{5}{3}\pi\right)$ (iii) $\frac{1}{32}\left(\frac{5}{2}\pi - \frac{22}{3}\right)$ (iv) $\frac{53}{480}$ (v) $\frac{49}{160}\sqrt{3}$ (vi) $\frac{4}{15}\sqrt{2}$ (vii) $\frac{5}{32}\pi$ (viii) $\frac{5}{32}\pi$ (ix) $\frac{1}{8}\left(-\frac{7}{8}\sqrt{3} + \frac{\pi}{2}\right)$ (x) $\frac{1}{16}\left(\frac{9}{4}\sqrt{3} + \pi\right)$ (xi) $\frac{1}{16}\left(-4 + \frac{3}{2}\pi\right)$.

Exercise 1.15 The proof goes almost verbatim to that of Example 1.28. We integrate the function $f(z) = \frac{z^2}{z^4+1}e^{iaz}$ along the closed curve C consisting of the line segement and the upper semi-circle with center at the origin and radius R. We put

$$\alpha = e^{\frac{2\pi i}{8}} = e^{\frac{\pi i}{4}} = \frac{1+i}{\sqrt{2}}, \; \beta = e^{\frac{3\pi i}{4}} = \frac{-1+i}{\sqrt{2}}.$$

$$\operatorname*{Res}_{z=\alpha} f(z) = \lim_{z\to\alpha} \frac{z^2}{(z-\bar{\alpha})(z^2+\sqrt{2}z+1)}e^{iaz} \quad (1.159)$$
$$= \frac{\alpha^2}{(\alpha-\bar{\alpha})(2\sqrt{2}\alpha)}e^{ia\alpha} = \frac{1}{4\sqrt{2}}e^{-\frac{a}{\sqrt{2}}}(1-i)e^{\frac{a}{\sqrt{2}}i}$$

and

$$\operatorname*{Res}_{z=\beta} f(z) = \lim_{z\to\beta} \frac{z^2}{(z-\bar{\beta})(z^2-\sqrt{2}z+1)}e^{iaz} \quad (1.160)$$
$$= \frac{\beta^2}{(\beta-\bar{\beta})(-2\sqrt{2}\beta)}e^{ia\beta} = \frac{1}{4\sqrt{2}}e^{-\frac{a}{\sqrt{2}}}(-1-i)e^{-\frac{a}{\sqrt{2}}i}.$$

Hence

$$\sum_{z=\alpha,\beta} \operatorname{Res} f(z) = \frac{1}{4\sqrt{2}} e^{-\frac{a}{\sqrt{2}}}(-2i)\left(\cos\frac{a}{\sqrt{2}} - \sin\frac{a}{\sqrt{2}}\right) \tag{1.161}$$

$$= \frac{-i}{2\sqrt{2}} e^{-\frac{a}{\sqrt{2}}}\left(\cos\frac{a}{\sqrt{2}} - \sin\frac{a}{\sqrt{2}}\right) = \frac{-i}{2} e^{-\frac{a}{\sqrt{2}}} \cos\left(\frac{a}{\sqrt{2}} + \frac{\pi}{4}\right).$$

Exercise 1.16 We treat (iii) only and prove (ii) at the same time. For this we integrate the function $f(z) = \frac{\log^3 z}{z^2+1}$ along the contour in Example 1.28, i.e. the contour C' consisting of the upper semi-circle C_R of radius R going to infinity, the line segment $[-R, -r]$, the *upper* semi-circle γ_0' of radius $r \to 0+$ and the line segment $[r, R]$. The only pole in the interior of C' is $z = i = e^{\frac{\pi}{2}i}$ and the residue of f is $-\frac{\pi^3}{16}$. Hence on one hand,

$$\int_{C'} f(z)\, \mathrm{d}z = -\frac{\pi^4}{8} i. \tag{A.20}$$

The integral along the line segment $[r, R]$ is $\int_r^R \frac{\log^3 x}{x^2+1}\, \mathrm{d}x$ while that along $[-R, -r]$ is $\int_r^R \frac{(\log x + \pi i)^3}{x^2+1}\, \mathrm{d}x$ in view of the increase of the argument along γ_0'. The latter approaches to

$$\int_0^\infty \frac{\log^3 x}{x^2+1}\, \mathrm{d}x - 3\pi^2 \int_0^\infty \frac{\log x}{x^2+1}\, \mathrm{d}x$$
$$+ i\left(3\pi \int_0^\infty \frac{\log^2 x}{x^2+1}\, \mathrm{d}x - \pi^3 \int_0^\infty \frac{1}{x^2+1}\, \mathrm{d}x\right).$$

Since $\int_0^\infty \frac{1}{x^2+1}\, \mathrm{d}x = \arctan \infty = \frac{\pi}{2}$, we have on the other hand

$$2\int_0^\infty \frac{\log^3 x}{x^2+1}\, \mathrm{d}x - 3\pi^2 \int_0^\infty \frac{\log x}{x^2+1}\, \mathrm{d}x + i\left(3\pi \int_0^\infty \frac{\log^2 x}{x^2+1}\, \mathrm{d}x - \frac{\pi^4}{2}\right).$$

This being equal to (A.20), we obtain on comparing the real and imaginary parts,

$$2\int_0^\infty \frac{\log^3 x}{x^2+1}\, \mathrm{d}x - 3\pi^2 \int_0^\infty \frac{\log x}{x^2+1}\, \mathrm{d}x = 0,\ 3\pi \int_0^\infty \frac{\log^2 x}{x^2+1}\, \mathrm{d}x - \frac{\pi^4}{2} = -\frac{\pi^4}{8}$$

or

$$\int_0^\infty \frac{\log^3 x}{x^2+1}\, \mathrm{d}x = 0, \quad \int_0^\infty \frac{\log^2 x}{x^2+1}\, \mathrm{d}x = \frac{\pi^3}{8} \tag{A.21}$$

on using $\int_0^\infty \frac{\log x}{x^2+1}\, \mathrm{d}x = 0$ which is due to Euler and established in Example 2.5, (i). It can be proved in verbatim to the above proof.

Use is made of the limit which is similar to the one that appeared in (2.43). Indeed, after simple estimation, we find that we need to show that

$\frac{r\log^3|r|}{1+r^2} \le r\log^3|r|$ tends to 0 as $r \to 0+$. An application of L'Hospital's rule proves that this limit and *a fortiori* the integral along γ_0' tends to 0. That the integral along the upper semi-circle C_R tends to 0 can be checked as in other cases.

The treatment of (iv) and (v) are almost word-for-word to the above case. For (v) we consider the integral $f(z) = \frac{\log^2 z}{(z^2+1)^2}$ along the curve C' above. Since

$$\mathrm{Res}_{z=i} f(z) = \lim_{z\to i} \frac{\mathrm{d}}{\mathrm{d}z}\left(\frac{\log^2 z}{(z+i)^2}\right) = \lim_{z\to i}\left(\frac{\frac{z+i}{z} - \log z}{(z+i)^3} 2\log z\right) = \frac{2-\frac{\pi}{2}i}{-8}\pi,$$

the integral is $-\frac{\pi^3}{8} - \frac{\pi^2}{2}i$. On the other hand,

$$\int_0^\infty \frac{(\log x + \pi i)^2}{(x^2+1)^2}\,\mathrm{d}x = \int_0^\infty \frac{\log^2 x - \pi^2}{(x^2+1)^2}\,\mathrm{d}x + 2\pi i\int_0^\infty \frac{\log x}{(x^2+1)^2}\,\mathrm{d}x,$$

so that

$$\int_0^\infty \frac{2\log^2 x - \pi^2}{(x^2+1)^2}\,\mathrm{d}x + 2\pi i\int_0^\infty \frac{\log x}{(x^2+1)^2}\,\mathrm{d}x = -\frac{\pi^3}{8} - \frac{\pi^2}{2}i.$$

Comparing the real and imaginary parts and noting that $\int_0^\infty \frac{1}{(x^2+1)^2}\,\mathrm{d}x = \frac{\pi}{4}$, we conclude that

$$2\int_0^\infty \frac{\log^2 x}{(x^2+1)^2}\,\mathrm{d}x - \frac{\pi^3}{4} = -\frac{\pi^3}{8} \tag{A.22}$$

$$2\pi i\int_0^\infty \frac{\log x}{(x^2+1)^2}\,\mathrm{d}x = -\frac{\pi^2}{2}i.$$

From this we confirm the evaluations given.

Exercise 1.21 We treat (ii) only. We apply Theorem 1.22 with $f(z) = \frac{z^a}{(z^2+b^2)^2}$. The first factor amounts to $-\frac{\pi}{\sin\pi a}e^{-\pi i a}$. Since $\mathrm{Res}_{z=bi} f(z) = i\frac{a-1}{4}b^{a-3}e^{\frac{\pi i}{2}a}$, the sum of residues is $i\frac{a-1}{4}b^{a-3}e^{\frac{\pi i}{2}a} - i\frac{a-1}{4}b^{a-3}e^{\frac{3\pi i}{2}a}$, it follows that the required integral is

$$\frac{\pi}{4}\frac{1}{\sin\pi a}i\left(e^{-\frac{\pi}{2}ia} - e^{\frac{\pi}{2}ia}\right)(1-a)b^{a-3} = \frac{\pi}{4}\frac{1}{\sin\pi a}2\sin\frac{\pi}{2}a(1-a)b^{a-3}$$

whence the result follows.

Exercise 2.36 We put

$$g(t) = \sum_{k=0}^{n-1} \frac{f^{(k)}(t)}{k!}(x-t)^k + A(x-t)^p, \tag{A.23}$$

where $A = A(x)$ is a function in x and so $g(x) = f(x)$.

We determine A so that

$$g(x_0) = f(x),$$

i.e.

$$A = A(x) = (x-t)^{-p}\left(f(x) - \sum_{k=0}^{n-1}\frac{f^{(k)}(t)}{k!}(x-t)^k\right). \tag{A.24}$$

Since $g(x) = g(x_0)$, it follows from Rolle's theorem that there is an x_1 between x and X_0 such that

$$\begin{aligned} 0 &= g'(x_1) \\ &= \sum_{k=0}^{n-1}\frac{f^{(k+1)}(x_1)}{k!}(x-x_1)^k - \sum_{k=1}^{n-1}\frac{f^{(k)}(x_1)}{k!}k(x-x_1)^{k-1} - p(x-x_1)^{p-1}A \\ &= \frac{f^{(n)}(x_1)}{(n-1)!}(x-x_1)^{n-1} - p(x-x_1)^{p-1}A, \end{aligned} \tag{A.25}$$

whence $A = \frac{f^{(n)}(x_1)}{(n-1)!p}(x-x_1)^{n-p}$ and so $R_n = A(x-t)^p$ is seen to be in the form as given.

Exercise 2.51 In (iii)-(vi), we put the first two or three arctan values as α, β, γ. (iii)

$$\alpha = \arctan\frac{1}{7}, \beta = \arctan\frac{3}{79} \tag{A.26}$$

We use the duplication formula $\tan 2\theta = \frac{2\tan\theta}{1-\tan^2\theta}$. Hence $\tan 4\alpha = \frac{14\cdot 24}{17\cdot 31}$ and $\tan 5\alpha = \tan(4\alpha+\alpha) = \frac{2879}{7\cdot 479}$. Using this and $\tan 2\beta = \frac{3\cdot 79}{76\cdot 41}$ to conclude $\tan(5\alpha+2\beta) = 1$.

(iv) $\tan 4\alpha = \frac{120}{7\cdot 17}$ and $\tan\left(4\alpha - \frac{\pi}{4}\right) = \frac{1}{239}$.

(v) $\tan 3\alpha = \frac{191}{488}$ and $\tan(3\alpha+\beta) = \frac{7}{17}$. Hence $\tan 2(3\alpha+\beta) = \frac{119}{120} = \tan\left(\frac{\pi}{4} - \gamma\right)$.

(vi) $\tan 4\alpha = \frac{120}{7\cdot 17}$ and $\tan(4\alpha - \beta) = \frac{8281}{8450}$. Hence $\tan(4\alpha+\beta+\gamma) = 1$.

Exercise 2.53 Let

$$\alpha = \arctan x, \beta = \arctan y \Leftrightarrow \tan\alpha = x, \tan\beta = y, -\frac{\pi}{2} < \alpha, \beta < \frac{\pi}{2}. \tag{A.27}$$

Then by (27.2),

$$\tan(\alpha+\beta) = \frac{x+y}{1-xy}, \quad -\pi < \alpha+\beta < \pi \tag{A.28}$$

provided that $xy \neq 1$ and $\alpha + \beta \neq \pm\frac{\pi}{2}$. To solve (A.28) in $\alpha + \beta$ we need to distinguish three cases. If $-\frac{\pi}{2} < \alpha, \beta < \frac{\pi}{2}$, then we obtain (2.185). In other two cases, we use other branches of the arctangent function.

E.g. in Exercise 2.50, with $x = \frac{1}{2}$, $y = \frac{1}{3}$ we have $\tan(\alpha + \beta) = 1$, $-\pi < \alpha + \beta < \pi$, whence (2.184).

For Exercise 2.51, (i), with $x = 2$, $y = 3$, we have $\tan(\alpha + \beta) = -1$ and $-\pi < \alpha + \beta < \pi$, whence $\arctan 2 + \arctan 3 = \frac{3}{4}\pi$. Exercise 2.52 also follows from Exercise 2.53 with $x = \frac{1}{p}$ and $y = -\frac{q}{p^2+pq+1}$.

Exercise 2.49 is an extremal case of Exercise 2.53 and may be thought of as a special case. Thus (2.185) covers all the cases with the understanding that the $xy = 1$ case means that the right-hand side is ∞, where the range of $\alpha + \beta$ is to be considered.

Bibliography

Ahlfors, L. (1979). *Complex analysis*, 3rd ed., MacGraw-Hill Book Co., New York.

Apostol, T. M. (1957). *Mathematical analysis*, Addison-Wesley, Reading.

Ban, N., Ogawa, H., Ono, M. and Ishida, Y. (2009). A servo control system using the loop shaping procedure, *World Academy of Sci., Engrg and Techn.* **60**, 150-153.

Brigham, E. O. (1974). *The fast Fourier transform*, Prentice-Hall, New Jersey.

Carlson, L. (1966). On convergence and growth of partial sums of Fourier series, *Acta Math.* **116**, 135-157.

Chakraborty, K., Kanemitsu, S. and Tsukada, H. (2009). *Vistas of special functions* II, World Scientific. London-Singapore etc.

Chakraborty K., Kanemitsu S., Kumagai, H. and Sato, K. (2009). Shapes of objects and the golden ratio, *J. Shangluo Univ.* **23**, 18-27.

Derrick, W. R. (1984). *Complex analysis and applications*, The Wadsworth math. series.

Dettman, J. W. (2012). *Applied complex variables*, Dover, New York 2012 (Reprint, first published by Macmillan in 1965).

Dienes, P. (1931). *The Taylor series: An introduction to the theory of functions of a complex variable*, OUP, London.

Donoghue Jr., W. J. (1969). *Distributions and Fourier transforms*, Academic Press, New York-London.

Doyle, J. C. and Stein, G. (1981). Multivariable feedback design: Concepts for classical/modern synthesis *IEEE Trans. Automatic Control* **26**, 4-16.

The architect Gaudi: his words, Chuo Koron Bijutsu Shuppan, Tokyo.

Gel'fond, A. O. (1971). *Residues and their applications*, Mir Publishers, Moscow.

Giardina, Ch. R. (1984). Band-limited signal extrapolation by truncated Bernstein polynomials, *J. Math. Anal. Appl.* **104**, 264-273.

Gradsteyn, I. S. and Ryzhik, I. M. (1980). *Tables of integrals, series, and products*, Academic Press, New York.

Grodins, F. S. (1963). *Control theory and biological systems*, Columbia Univ. Press, New York and London.

Hirano, G., Takahashi, K., Kaida, T., Kanemitsu, S. and Matsuzaki, T. (2012). Legitimation of the use of fancy tools, *Kayanomori* **16**, 8-14.

Hunt, R. (1968). On the convergence of Fourier series, *Orthogonal expansions and their continuous analogues Proc. Conf., Edwardsville*, III, *Carbondale* III., *Southern Illinois Univ. Press*, 235-255.

Imai, I. (1963). *Fluid mechanics*, Science-sha, Tokyo (in Japanese).

Ito, H. (1968). *Mathematics of* e, Kouseisha, Tokyo,.

Ivanov, V. I. and Trubetskov, M. K. (1995). *Handbook of conformal mapping with computer-aided visualization*, CRC Press, Boca Raton etc.

Kanemitsu, S. and Tsukada, H. (2007). *Vistas of special functions*, World Scientific, Singapore etc.

Kanemitsu, S. and Tsukada, H. (2014). *Contributions to the theory of zeta functions the modular relation supremacy*, World Sci. London etc.

Karatsuba, A. A. (1995). *Complex analysis in number theory*, CRC Press, Boca Raton etc.

Kimura, H. (1984). Robust stabilizability for a class of transfer functions, *IEEE Trans. Automatic Control* **27**, 788-793.

Kimura, H. (1997). *Chain scattering approach to* H^∞*-control*, Birkhäuser, Boston-Basel-Berlin.

Kumagai, N. (2007). *Ciphers in the times of the Web–Challenge of net security*, Chikuma-shobo, Tokyo.

Kuroš, A. G. (1974). *The theory of groups*, Vol. II, AMS-Chelsea, Providence, R.I.

Lang, S. (2012). *Algebraic number theory*, Addison-Wesley, New York.

Li, F.-H., Wang, N.-L. and Kanemitsu, S. (2012). *Number theory and its applications*, World Sci., London etc.

Lindelöf, E. (1905). *Calcul des résidus et ses applications a la théorie de fonctions*, Gauthier-Villars, Paris.

Lorentz, G. G. (2012). *Bernstein polynomials*, Amer. Math. Soc., Providence R.I.

MacFarlane, D. and Glover, K. (1996). A loop shaping design procedure using H control, *IEEE Trans. Automatic Control* **37**, 491-499.

Marshall, K. E. and Stray, A. (1992), Interpolating Blaschke products, *Pacific J. Math.* **173**, 759-769.

Miller, D. S. (1970). *Advanced complex calculus*, Dover, New York.

Mitrinović, D. S. and Kečkić, J. D. (1984). *The Cauchy method of residues Theory and Applications*, Reidel Publ., Dordrecht etc.

Papoulis, A. (1962). *The Fourier integral and its applications*, McGraw-Hill.

Rademacher, H. (1973). *Topics in analytic number theory*, Springer Verl, Berlin-Heidelberg.

Ruderman, D. L. and Bialek, W. (1999). Seeing beyond the Nyquist limit, *in Neural codes and distributed representations–Foundations of neural computation ed. by L. Abbott and T. J. Sejmowski MIT Press, London*, 119-127.

Rudin, W. (1986). *Real and complex analysis*, MacGraw-Hill, New York etc.

Schmidt, E. (1904). *Die Cauchy'sche Methode der Auswertung bestimmter Integrale zwishcen reellen Grenzen*, Hofbuchdruckerei zu Gutenberg Carl Grüninger, Stuttgart.

Segal, S. L. (1981). *Nine introductions in complex analysis*, North-Holland, Amsterdam etc.

Shannon, Cl. (1949). Communication in the presence of noise, *Proc. IRE* **37**, 10-21.

Splettstösser, W. (1985). Some aspects on the reconstruction of sampled signal functions, *the Road-Vehicle-System and Related Mathematics, H. Neunzert (ed.), B.G.Teubner,*, 126-142.

Srivastava, H. M. and Choi, J. -S. (2001). *Series associated with the Zeta and related functions*, Kluwer Academic Publishers, Dordrecht etc.

Takada, T. (1985). *Introduction to the theory of Schwarz distribution*, Nihon-hyoron-sha, Tokyo.

Takahashi, K., Hirano, G., Kaida, T., Kanemitsu, S., Tsukada, H. and Matsuzaki, T. (2011). Record of the second and the third interdisciplinary seminars, *Kayanomori* **14**, 64-72.

Takahashi, K., Hirano, G., Kaida, T., Kanemitsu, S. and Matsuzaki, T. (2011). On linear recurrences and their applications, *Kayanomori* **15**, 13-20.

Takahashi, K., Hirano, G., Kaid,a T., Kanemitsu, S. and Matsuzaki, T. (2012). The Cool'n Tacky al-Khwārizmī, *Kayanomori* **17**, 17-22.

Takahashi, K., Matsuzaki, T. , Hirano, G. , Kaida, T., Fujio, M., and Kanemitsu, S. (2014). Restoration and extrapolation of band-limited signals, *Kayanomori* **20**, 10-14.

Takahashi, K., Matsuzaki, T. , Hirano, G. , Kaida, T., Fujio, M., and Kanemitsu, S. (2014). Fluctuations in science and music, *Kayanomori* **21**, 1-7.

Tatuzawa, T. (1980). *Theory of functions*, Kyoritsu-shuppan, Tokyo.

Titchmarsh, E. C. (1939). *The theory of functions*, 2nd ed. Oxford UP, London.

Vidyasagar, M. and Kimura, H. (1983). Robust controllers for uncertain linear multivariable systems, *Automatica* **22**, 585-601.

Weaver, H. J. (1983). *Applications of discrete and continuous Fourier transforms*, Wiley, New York etc.

Whittaker, E. T. (1915), On the functions which are represented by the expansions of the interpolation theory, *Proc. Royal Soc. Edinburgh* **35**, 181-194.

Whittaker, E. T. and Watson, G. N. (1927). *A course of modern analysis*, 4th ed., Cambridge UP, Cambridge.

Woodward, P. M. (1953). *Probability and information theory, with applications to radar*, Pergamon Press, London.

Zames, G. and Francis, B. A. (1992). Feedback minimax sensitivity, and optimal robustness *IEEE Trans. Automatic Control* **28**, 759-769.

Zou, Z. -G. Robust controller for servo systems, *unpublished.*

Index

Printed in the United States
By Bookmasters